72nd Northeastern Weed Science Society Annual Meeting 2018

Proceedings – Volume 72

Philadelphia, Pennsylvania, USA
9 – 11 January 2018

ISBN: 978-1-5108-6295-1

Printed from e-media with permission by:

Curran Associates, Inc.
57 Morehouse Lane
Red Hook, NY 12571

CURRAN ASSOCIATES INC. proceedings.com

Some format issues inherent in the e-media version may also appear in this print version.

Printed by Curran Associates, Inc. (2018)

Northeastern Weed Science Society (NEWSS)
P.O. Box 25
Woodstown, NJ 08098
USA

Phone: 315-209-7580

northeasternweedscience@hotmail.com

Additional copies of this publication are available from:

Curran Associates, Inc.
57 Morehouse Lane
Red Hook, NY 12571 USA
Phone: 845-758-0400
Fax: 845-758-2633
Email: curran@proceedings.com
Web: www.proceedings.com

PROCEEDINGS - VOLUME 72

72nd Annual Meeting of the

Northeastern Weed Science Society

January 9-11, 2018 – Hilton Philadelphia Penn's Landing - Philadelphia, PA

Site of 2019 Joint Meeting*

January 8-10, 2019

Delta Hotels Baltimore Hunt Valley
245 Shawan Road
Hunt Valley, MD

Participating Societies:
Northeastern Weed Science Society
American Society of Agronomy – Northeastern Regional Branch
American Society of Horticultural Science – Northeast Division

To order a copy of the Proceedings contact:
CURRAN ASSOCIATES, INC
57 Morehouse Lane, Red Hook, NY 12571
Ph: 1 845 758 0400 x112

§ Cover: The winning image in the 2018 NEWSS Photo Contest, submitted by Thierry Besançon, Rutgers University.

Proceedings

of the

Seventy Second Annual Meeting

of the

Northeastern Weed Science Society

Michael L Flessner, Editor

NORTHEASTERN WEED SCIENCE SOCIETY

2018 SUSTAINING MEMBERS

Platinum

BASF The Chemical Company

syngenta

Gold

FMC

Bayer CropScience

MONSANTO

Silver

Gowan USA The Go To Company

ohp Partners with solutions

UPI

GDM

For-Shore Weed Control Lawn Care Tree & Shrub Care

Bronze

LABServices

ACDS Research, Inc.

Table of Contents

NEWSS EXECUTIVE COMMITTEE – OFFICERS

President	**Carroll Moseley** Syngenta Carroll.moseley@syngenta.com
President-elect	**Dan Kunkel** Rutgers University kunkel@aesop.rutgers.edu
Vice President	**Art Gover** The Pennsylvania State University aeg2@psu.edu
Treasurer	**Kurt Vollmer** University of Delaware kvollmer@udel.edu
Past President	**Randy Prostak** University of Massachusetts rprostak@umext.umass.edu

NEWSS EXECUTIVE COMMITTEE – MEMBERS

Membership Chair	**Stacey Pickard** ACDS Research, Inc. smpickard@acdsresearch.com
Editor	**Michael Flessner** Virginia Tech flessner@vt.edu
Public Relations	**Katelyn Venner** BASF katelyn.venner@basf.com
Research and Education Coordinator	**Quintin Johnson** University of Delaware quintin@udel.edu
Sustaining Membership	**Wesley Everman** North Carolina State University wes_everman@ncsu.edu
CAST Representative	**Rob Richardson** North Carolina State University rob_richardson@ncsu.edu
Graduate Student Representative	**Theresa Reinhardt** North Carolina State University tareinha@ncsu.edu
WSSA Representative	**Rakesh Chandran** West Virginia University rschandran@mail.wvu.edu
Science Policy Director	**Lee Van Wychen** Weed Science Society of America lee.vanwychen@wssa.net

2019 NEWSS MEETING PROGRAM COMMITTEES

Program Committee	Chair:	Dan Kunkel, Rutgers University Art Gover, The Pennsylvania State University
Agronomy	Chair:	Larissa Smith, Syngenta
Vegetation Management and Restoration	Chair: Chair elect:	Lane Heimer, Maryland Dept. of Agriculture Art Gover
Turfgrass and Ornamentals	Chair: Chair elect:	Matt Elmore, Rutgers University Jatinder Aulakh, The Connecticut Agricultural Experiment Station
Research Posters	Chair:	Kate Venner, BASF
Vegetables and Fruit	Chair: Chair-elect:	Charlie Cahoon, North Carolina State University Ramón Arancibia, Virginia Tech

2018 NEWSS COMMITTEES

Student Paper Contest	Chair:	Randy Prostak, University of Massacusetts
Student Poster Contest	Chair:	Matt Elmore, Rutgers University
Photo Judging Contest	Chair:	Hilary Sandler, University of Massacusetts
Site Selection Committee	Chair: Chair-elect:	Art Gover, The Pennsylvania State University Dan Kunkel, Rutgers University
Awards Committee	Chair:	Randy Prostak, University of Massacusetts
Nomination Committee	Chair:	Randy Prostak, University of Massacusetts
Resolutions Committee	Chair:	Jerry Baron, Rutgers University
Collegiate Weed Contest	Chair:	Dan Kunkel, Rutgers University
Audit Committee	Chair:	Kurt Vollmer, University of Delaware
Archives Committee	Chair:	Dan Kunkel, Rutgers University
Weed Science Field Days	Chair:	Quintin Johnson, University of Delaware
Past Presidents Committee	Chair:	Randy Prostak, University of Massacusetts
Herbicide Resistance	Chair:	Mark VanGessel, University of Delaware
Endowment Committee	Chair:	Art Gover, The Pennsylvania State University

2018 AWARDS, NEWSS ANNUAL MEETING

NEWSS Fellow Award – Mark VanGessel

Award of Merit - Renee Keese

Outstanding Researcher Award - Antonio DiTommaso

Outstanding Educator Award - Rob Richardson

Robert D. Sweet Outstanding Ph.D. Student Award - Sandeep Rana

M. Garry Schnappinger Service Recognition Award - José Javier Vargas Almodóvar

NEWSS CONTEST – WINNERS, 2018

STUDENT PAPER

1st place paper – Elizabeth Buck, University of Guelph, "Investigating Herbicide Uses to Further Interseeded Cereal Rye (Secale cereale) Use in Vegetable Systems"

2nd place paper – Brandon Schrage, North Carolina State University, "Influence of Timing and Tank Mixtures on Control of Multiple-resistent Common Ragweed (*Ambrosia artemisiifolia*)"

3rd place paper – J. Harrison Ferebee IV, Virginia Tech, "Evaluating Preemergence Herbicides in Pumpkins"

Honorable mention – Jordan Craft, Virginia Tech, "Mechanical and Chemical Programs for Bermudagrass Control in Creeping Bentgrass"

STUDENT POSTER

1st place: Eugene Law, Cornell University, "Comparing Weed Communities in Annual and Perennial Small Grain Cropping Systems during the Establishment Year"

2nd place: Brandon Schrage, North Carolina State University, "Identification of Cross- and Muliple-resistance in Ambrosia artemisiifolia in North Carolina"

3rd place: Lucas Rector, Virginia Tech, "Comparison of Dicamba- versus Glufosinate-based Herbicide Programs in Soybeans"

PHOTO CONTEST

1st place photo – Thierry Besançon, Rutgers University, Bloom of Carolina redroot

2nd place photo – Eugene Law, Cornell University, Blooming common teasel

3rd place photo – Sandeep Rana, Monsanto, Panoramic view of a mustard field

2018 NEWSS COLLEGIATE WEED SCIENCE CONTEST

Hosted by ACDS Research, Inc., North Rose NY

(Photos courtesy of Shawn Askew and Katelyn Venner)

The 2017 NEWSS Annual Weed Contest was hosted at ACDS Research, Inc. on July 25, 2017. Twelve graduate teams and nine undergraduate teams compete for a total of 79 students from six universities across the region. The contest went smoothly and despite the cool, wet conditions. The contest site was beautiful, and allowed students to see how crops like onions, radishes and cabbage are grown in Upstate New York. The contest concluded with the awards ceremony at the beautiful Webster Arboretum. The contest was quite competitive, Virginia Tech Graduate Team 1 won the graduate student division and University of Guelph Undergraduate Team 1 won the undergraduate division. Grant Jordan and the team at ACDS stepped up this year and hosted an amazing contest on very short notice. Many thanks to them! It could not have been possible without financial support and many volunteers.

Farmer Problems

At the Arboretum

Webster Arboretum, Webster NY

Dwight and the Penn State crew enjoy each other's company over drinks outside the banquet hall.

The ACDS Research, Inc. team photo. Many thanks to these folks for putting together a great contest!!

Fearless NEWSS Past-President Shawn Askew prepares to take the group photo from the best vantage point.

Awards

Undergraduate Team Awards

First Place Undergraduate

Guelph Undergraduate Team 1

(Alex Sanders, Todd Fre, Peter Buys, and Brett Hilker)

Second Place Undergraduate

Guelph Undergraduate Team 3

(Aniken Young, Julie French, Anita Rastapkevicius, and Emily Priester)

Third Place Undergraduate

Cornell Undergraduate Team 1

(Cynthia Sias Roxana Padilla, Elsa de Becker, and Patricia Chan)

Awards… cont'd

Undergraduate Individial Awards

First Place Undergraduate

Alexander Sanders (Guelph)

Second Place Undergraduate

Patricia Chan (Cornell)

Third Place Undergraduate

Todd Frey (Guelph)

Awards… cont'd

Graduate Team Awards

First Place Graduate Team

Virginia Tech Graduate Team 1

(Kara Pittman, Shawn Beam, John Brewer, and Camden Shelton)

Second Place Graduate Team

Guelph Graduate Team 1

(Meghan Grguric, Brittany Hedges, Elizabeth Buck, and Lauren Benoit)

Third Place Graduate Team

North Carolina State University Grad 1

(Brandon Schrage, John Sanders, Manish Bansal, and Cole Smith)

Awards... cont'd

Graduate Individual Awards

First Place Graduate Student

John Brewer (Virginia Tech)

Second Place Graduate Student

Brendan Metzger (Guelph)

Third Place Graduate Student

Shawn Beam (Virginia Tech)

HERBICIDE LONGEVITY IN CONTAINER SUBSTRATES. J. Aulakh *, USDA-ARS, Wooster, OH
(5)

ABSTRACT

[no abstract submitted]

CHANGE IN MICROBIAL COMMUNITIES OF PINE BARK SUBSTRATES OVER TIME. S. Valles*, J. Altland, and F. Michel, Ohio State University, Wooster, OH (6)

ABSTRACT

A large body of research has addressed the biological community in soilless substrates. Most of this research pertains to specific sets of pathogens or plant growth promoting microbes. Very little is known about the overall microbial community in terms of species range, diversity and relative population density. The objectives of this research were to analyze microbial community structure in a typical pine bark substrate used for nursery crop production and determine the impacts of compost amendment and plant growth on these communities. Three substrates (v/v); 80:20:0, 80:10:10 and 80:0:20, pine bark:sphagnum peat:leaf compost were prepared. The substrates were filled into 20 L nursery containers and half were planted with a single birch liner (*Betula nigra* 'Cully') from a 50-cell flat while the other half remained fallow. Containers were fertilized with 73 g controlled release fertilizer. There were six single-pot replications per treatment. The microbial consortia in the potting media were characterized using high-throughput ribosomal RNA gene and intergenic spacer region sequencing. Representative samples (500 g) were taken from each container starting on April 12 and monthly thereafter throughout the growing season (4 months) and stored at -22 °C until analyzed. DNA was extracted and purified using DNeasy PowerSoil Kit components. The product size was verified by gel eletrophoresis. A 25 µL aliquot at a 5 $mg \cdot mL^{-1}$ concentration was used for PCR amplification. Universal as well as population-specific bacterial and fungal primers were used to identify and quantify tens of thousands of individual ribotypes within each sample by comparison of the amplified sequences to 16S gene and ITS databases. The data was processed using an open-source bioinformatics pipeline (QIIME). Bacterial communities of the substrates immediately after potting differed in composition. The compost amended substrate (80:0:20) was dominated by Proteobacteria (37.4%), Actinobacteria (35.6%) and Acidobacteria (23.0%). The peat amended substrate was initially dominated by Proteobacteria but also had relatively large percentages of Chloroflexi and Bacteroidetes. Over time, Bacteroidetes increased while Actinomycetes and Acidobacteria decreased in all of the mixes. While there were initially differences in microbial communities between the substrate types, after 2 months the communities in all substrates were similar. Planting trees or adding compost to the media did not have a strong impact on bacterial community composition after 2 months.

EFFICACY OF TAVIUM HERBICIDE PLUS VAPORGRIP TECHNOLOGY IN DICAMBA TOLERANT SOYBEANS AND COTTON. E.M. Hitchner*, D.J. Porter, S. Payne, J.C. Holloway, D.L. Bowers, and B.R. Miller, Syngenta, NJ (7)

ABSTRACT

Tavium Plus VaporGrip Technology is a new herbicide under development by Syngenta for use in dicamba tolerant soybeans and cotton. It is a convenient premix containing three key components: dicamba, a Group 4 herbicide, *S*-metolachlor, a Group 15 herbicide, and VaporGrip Technology which decreases the volatility of dicamba and reduces the chance for off-site movement. Tavium Plus VaporGrip Technology provides postemergence control of over 50 broadleaf weeds as well as extended residual control of key broadleaf weeds such as waterhemp and Palmer amaranth as well as troublesome grasses. Tavium Plus VaporGrip Technology offers flexibility in application timing by allowing one application from preplant burndown through preemergence and one application postemergence in both dicamba tolerant cotton and soybeans. By employing two modes of action, Tavium Plus VaporGrip Technology is an effective resistance management tool which will fit well into an integrated weed management program by delivering postemergence control and enabling overlapping residual activity.

PRELIMINARY STUDIES ON NOSTOC CONTROL IN NURSERY SITES. C. Harlow* and J. Neal,
North Carolina State University, Raleigh, NC (9)

ABSTRACT

Preliminary studies were conducted 2015 - 2017 to evaluate effectiveness of preemergence herbicides and herbicide combinations in controlling Nostoc in container nursery gravel sites. During summer 2015, indaziflam, flumioxazin, diuron, and simazine, alone or in combination with diquat, were applied to gravel roadways at a local nursery. Diquat alone also was applied. The roadways were scraped with a box-blade prior to treatments. Nostoc growth over time was monitored, and % Nostoc cover in plots was recorded. Diuron and simazine treatments showed reduced Nostoc growth compared to nontreated plots during the experiment. The addition of diquat did not affect Nostoc growth. A similar experiment was conducted in spring 2017 at the same nursery. For that experiment, treatments included four rates of simazine (0.25, 0.5, 1, or 2 qt/A) and four rates of diuron (0.25, 0.5, 1, or 2 qt/A). Roadways were not scraped prior to treatment. In that experiment, no treatment reduced Nostoc cover relative to the nontreated. A third experiment was initiated summer 2017. Treatments included three rates of simazine (0.25, 0.5, or 1 qt/A) and three rates of diuron (0.25, 0.5, or 1 qt/A). Additionally, half of the plots were scraped with a box-blade prior to treatment and half were not disturbed, in order to evaluate the effectiveness of this technique. Diuron at 0.5 and 1 qt/A reduced Nostoc cover through 45 days after treatment (DAT) in plots that had not been scraped. By 76DAT, no differences in cover were observed in any plot. In plots that had been scraped prior to treatment, all treatments reduced Nostoc cover relative to the nontreated through 45DAT. By 76DAT, simazine at 0.25 and 0.5 qt/A no longer showed reduced cover; reduced cover was observed in all other treatments. This study is ongoing. Future experiments will further evaluate scraping as a control tool and also more clearly define required herbicide dosages.

TOWER +/- PENDULUM EC AND FREEHAND EFFICACY ON COMMON WEEDS OF CONTAINER NURSERIES. J. Neal* and C. Harlow, North Carolina State University, Raleigh, NC (10)

ABSTRACT

An experiment was conducted to evaluate the response of common nursery weeds to dimethenamid-P herbicides (Freehand and Tower) applied within the labeled dose range. An additional objective was to compare the spray-applied dimethenamid-P alone or tank mixed with pendimethalin EC. The experiment was conducted at the Horticulture Crops Research Station at Castle Hayne, NC. On April 18, 2017 nursery trade 2-gallon pots (7 L) were filled with hammer milled pine bark amended with slow release fertilizer. Pots were watered to settle the substrate; one day later herbicide treatments were applied. Treatments included Freehand 1.75G at 150, 175 and 200 lb/A, Tower 6EC at 21, 27 and 32 oz/A alone or combined with Pendulum EC at 3.6 pt/A. These treatments were compared to a non-treated control and three industry-standard herbicides. Granular herbicides were applied in pre-weighed aliquots using a handheld shaker jar. Spray applications were made with a CO_2 pressurized 3-L bottle sprayer equipped with two 8004 flat fan nozzles and calibrated to deliver 30 GPA. Treatments were arranged in a randomized complete block design with 4 replicates. After treatment, pots received about 0.5 in. of overhead irrigation and thereafter received about 0.6 inch of overhead irrigation daily. April 20, 2017 (one day after treatment) pots were surface seeded with weeds, about 30 to 40 viable seeds per pot, 3 pots (=3 subsamples) of each species per experimental unit. Weed species were: large crabgrass, dogfennel, eclipta, longstalked phyllanthus, spotted spurge, and bittercress. Weed control was visually evaluated 4 & 6 weeks after initial application (WAA1). Weed seedling emergence was counted about 4 WAA1. Following the 6 WAA1ratings, all pots were treated with glyphosate, hand weeded 8WAA1 then re-treated and re-seeded. Visual evaluations continued 4, 6, and 8 WAA2. Above-ground fresh weights were recorded 8WAT2. Increasing the Freehand dose from 150 to 200 lb/A did not improve control of weeds in this study except dogfennel that was better controlled with increasing doses. There were no differences between Freehand and Tower in control of bittercress, spurge, dogfennel or eclipta. However, Freehand provided better control of phyllanthus and crabgrass. The addition of pendimethalin EC to Tower treatments improved control of phyllanthus and crabgrass over Tower alone. For pendimethalin EC plus dimethanamid-P tank mixes, there were no dose responses for the weeds tested. These data suggest that for most common nursery weeds, applying Freehand at doses greater than 150 lb/A likely will not significantly improve weed control. Additionally, tank mixing pendimethalin with Tower will improve control of certain species, long stalked phyllanthus in particular, and may reduce the dose of Tower needed.

IMPACT OF DROUGHT ON COMPETITION BETWEEN IVYLEAF MORNINGGLORY AND CORN. A. DiTommaso*, K.H. Averill, S.H. Morris, and M.C. Hunter, Cornell University, Ithaca, NY (11)

ABSTRACT

Increasing duration of growing season drought associated with climate change could alter competitive interactions between crop and weed plants. We tested the effect of drought on ivyleaf morningglory (*Ipomoea hederacea* Jacq.) (IMG) competition in glyphosate-resistant corn (*Zea mays* L.) during the 2016 and 2017 growing seasons in Ithaca, NY. We created drought conditions (7 wk duration) with rainout (i.e., rain exclusion) shelters established after planting and early season management practices. Rainout shelters, each 2.7 m wide by 3 m long by 3 m tall, were fitted with gutters and drainage tubing to divert water away from plots. Corn was planted in late May at a density of 71,660 plants ha^{-1} with 76 cm row spacing. In early June, IMG seedlings (of similar size to corn seedlings) were transplanted into corn rows at five densities (0, 1, 2, 4, 8 plants m^{-2}). Weeds other than IMG were controlled through a combination of glyphosate applied once prior to transplanting IMG, inter-row cultivation in late June (2016 only), and weekly hand-weeding. Because of a record-dry June 2016, non-drought treatment plots were irrigated weekly with approximate 2.5 cm watering events before establishing rainout shelters in late July and until crop and weed harvest in mid-September. In 2017, no irrigation occurred and shelters were established slightly earlier (in mid-July) and harvest occurred two weeks earlier than in 2016. We measured soil volumetric water content at least once per week during the growing seasons using fixed-in-place, time-domain reflectometer probes. In 2016 soil water content was lower in drought (D) plots with rainout shelters (0.086 ± 0.003 cm^3 water cm^{-3} soil) compared with non-drought (ND) plots without rainout shelters (0.116 ± 0.003 cm^3 water cm^{-3} soil) ($p < 0.001$), as well as in the much wetter 2017 season (D: 0.180 ± 0.006 cm^3 water cm^{-3} soil; ND: 0.246 ± 0.007 cm^3 water cm^{-3} soil) ($p < 0.001$). We found no evidence that drought alters the competitive effect of IMG in corn silage, i.e., interactions between drought and IMG density were not statistically significant ($p > 0.05$). Furthermore, the drought treatment did not affect IMG biomass ($p = 0.8$). Nonetheless, our data show that the drought reduced fresh corn silage yield by 12 % in 2016 (D: 19,100 ± 900 kg ha^{-1}; ND: 21,600 ± 700 kg ha^{-1}) and by 8 % in 2017 (D: 28,200 ± 700 kg ha^{-1}; ND: 30,500 ± 900 kg ha^{-1}) in 2017 ($p = 0.001$). We found no relationship between IMG density and silage yield. These results suggest that, in central NY, drought during the growing season is unlikely to alter ivyleaf morningglory competitiveness in silage corn and this vine poses low risk for silage yield-losses regardless of drought.

EFFECT OF SORGOLEONE (SORGHUM ROOT EXUDATE) ON GROWTH OF DIFFERENT WHEAT VARIETIES. M. Bansal* and W.J. Everman, North Carolina State University, Raleigh, NC (12)

ABSTRACT

Sorghum production has gained interest in recent years as regional grain demands increased which lead swine producer to offer a competitive sorghum grain price. Sorghum can be a good alternative for corn in rotation with wheat. Sorghum has ability to tolerate hot dry weather, a condition that can be challenging for corn in drought season. However, with the advantages, sorghum has some disadvantages as well when used in rotation. Grain sorghum is known to have has negative impact on the following crop due to its allelochemical properties. Sorghum in known to produce allelochemical called 'Sorgoleone'. Plants can produce these chemicals either by roots when they are still alive or by dead decaying matter and are known to have negative impact on weeds and following crops. There are concerns about sorghum affecting following winter wheat growth when grown in rotation in North Carolina. Four different wheat varieties from each of four type of winter wheat; soft red, sort white, hard red, and hard white is selected to evaluate the impact of sorgoleone. Sorgoleone was mixed in agar and applied at 0 (control), 25, 50, 100, 150, 200, and 300 $\mu g\ ml^{-1}$ under controlled environmental conditions. A week after placing seeds on the petri dishes, growth will be measured in terms of shoot length. We expect that sorgoleone might have negative impact on growth of wheat.

EVALUATION AND DEMONSTRATION OF INTEGRATED DISEASE MANAGEMENT OPTIONS FOR ORGANIC TOMATO AND CUCUMBER PRODUCTION. D.E. Telenko* and D.G. Ludwig,

Cornell University, East Aurora, NY (14)

ABSTRACT

Late blight (*Phytophthora infestans*) and downy mildew (*Pseudopernospora cubensis*) have the potential to significantly reduce tomato (*Solanum lycopersicum*) and cucumber (*Cucumis sativus*) yields in New York State, especially for organic growers. Our objectives in these trials were: 1. Demonstrate the importance of using disease resistant cultivars. 2. Evaluate a new biopesticide for organic tomato and cucumber production. Organic fungicide programs and two cultivars each of tomato and cucumber were evaluated at the Cornell Lake Erie Research and Extension Laboratory (CLEREL), Portland, NY on a Chenagno gravelly loam soil. A split-plot design was used with three replications and treatments arranged in a randomized complete block design. Organic fungicide programs were the main plot and consisted of one 6.1 m long raised black plastic mulched bed subdivided into 1.5 m subplots for each of the two cultivars in both tomato and cucumber. Calendar sprays of each program were applied on 29 Jun, 13 Jul, 20 Jul, 2 Aug, 16 Aug, and 25 Aug and included 1) Untreated, 2) Standard copper spray (Badge X2) at 1.12 kg/ha, 3) Hydrogen dioxide (Oxidate 2.0) at 1.0%, 4) Standard copper spray (Badge X2) at 1.12 kg/ha + hydrogen dioxide at 0.5%, and 5) *Bacillus subtilis,* strain MBI600 (Serifel) at 280 g/ha in-furrow on 7 June; foliar spray on 29 Jun, 20 July, and 16 Aug; alternated with copper at 1.12 kg/ha spray on 13 July, 2 Aug, and 25 Aug. Tomato cultivars included Polbig (susceptible to late blight) and Iron Lady (resistant to late blight). Cucumber cultivars included Marketmore 76 (susceptible to downy mildew) and DMR 264 (resistant to downy mildew). Sprays provided protection, but no differences were detected between them. In both crops, the resistant variety had significantly less disease than the susceptible. Iron lady yielded twice as much as Polbig, while DMR 264 was slower to mature, therefore yields were significantly less than Marketmore 76. These trials demonstrated the importance of utilizing disease resistance when available in organic tomato and cucumber production.

EXAMINING SPECTRAL REFLECTANCE PROPERTIES OF ITALIAN RYEGRASS USING MULTISPECTRAL IMAGERY. J. Sanders* and W.J. Everman, North Carolina State University, Holly Springs, NC (15)

ABSTRACT

With the tremendous amount of innovation in the consumer unmanned aerial vehicle (UAV) market in recent years, UAVs could very well prove to be an adaptable and efficient means of scouting fields for weed infestations. Aerial imagery has long been used in agriculture to quantify such metrics as crop vigor and water content as well as to map disease and fertility issues, but due to cost and practicality issues surrounding traditional means of acquiring aerial imagery, the utilization of these technologies has been often limited. UAVs offer a relatively low cost and highly efficient means of acquiring remotely sensed aerial imagery for the monitoring of crop health and pest issues and for use as a decision aid in making management decisions. In 2016, field studies were established to study the effects of image acquisition date, altitude and weed density on the spectral reflectance properties of mixed stands of wheat and Italian ryegrass (*Lolium perenne* spp. *multiflorum*). 5-band multispectral imagery was collected from the sites throughout the season and was used to record measurements of spectral reflectance. Spectral readings were then subject to an analysis of variance to study how these factors influenced the spectral response both alone and in interactions with one another. Date effects were highly significant influencers of spectral reflectance, with reflectance of red, blue, green and red edge light generally decreasing throughout the season while near infrared reflectance gradually increased as the season progressed. The effects of altitude appear insignificant on their own, rather altitude effects on spectral reflectance would appear to be depend on conditions specific to the date of image acquisition. Finally, weed density significantly influenced spectral reflectance only within one band, suggesting this band's utility for discrimination between different weed population densities.

BACTERIA OF SMOOTH CRABGRASS (*DIGITARIA ISCHAEMUM*) SEED INHIBIT COMPETITOR PLANT SPECIES.
M.T. Elmore*, J.F. White, K.L. Kingsley, K.H. Diehl, and D.P. Tuck, Rutgers University, New Brunswick, NJ (16)

ABSTRACT

We hypothesized that bacteria associated with *D. ischaemum* and *P. annua* seeds would affect seedling growth and antagonize competitor plants. Bacteria and fungi associated with surface-sterilized *D. ischaemum* seeds and non-surface sterilized *P. annua* seeds were isolated for further study. Among twenty-four bacterial strains isolated and evaluated against *Taraxacum officinale* in seedling culture, ten were antagonistic and four of those were antagonistic enough to increase mortality. All four strains that increased mortality were isolated from *D. ischaemum* seeds. Of these, two were characterized as *Pantoea*spp. and evaluated in more detail alone and in combination with a *Curvularia* sp. also isolated from *D. ischaemum* seed. These bacteria caused >65% *Trifolium repens* seedling mortality but did not affect *P. annua* seedlings. Effects on *D. ischaemum* mortality were inconsistent. Whether alone or in combination with bacteria, *Curvularia* was highly pathogenic to *D. ischaemum* and *T. repens*. However, the *Curvularia* sp. increased the positive root gravitropic response of *P. annua* roots. These experiments demonstrate that bacteria associated with *D. ischaemum* seed may be antagonistic to competitor forbs. Future research should evaluate these bacteria in greenhouse culture and elucidate the mechanism by which they cause mortality.

IDENTIFICATION OF CROSS- AND MULTIPLE-RESISTANCE IN *AMBROSIA ARTEMISIIFOLIA* IN NORTH CAROLINA. B. Schrage*, W.J. Everman, J. Sanders, and T.N. OQuinn, North Carolina State University, Raleigh, NC (17)

ABSTRACT

Control of herbicide-resistant dicotyledonous weed species is one of the greatest obstacles to securing optimum soybean yields in North Carolina. From 2013 to 2015 growers from Northeastern counties reported increasing difficulty controlling common ragweed with diphenylethers and N-phenylphthalimides-many of which are labeled for *Ambrosia* control in soybeans. A common ragweed biotype was confirmed to be resistant to ALS-inhibitors, PPO-inhibitors and glyphosate in 2015. A bioassay was warranted in order to determine if common ragweed had developed cross-resistance across herbicide families within the respective herbicide mechanisms of action.

In 2017, a resistant population from Currituck County and two susceptible populations from Edgecombe County and Leeland, Mississippi were screened against glyphosate and several rates of PPO- and ALS-inhibiting herbicides. PPO-Inhibiting herbicides included fomesafen, carfentrazone, and flumiclorac which belong to the chemical familys diphenylether, aryl triazinone, and N-phenylphthalimide, respectively. ALS-Inhibiting herbicides included chlorimuron, cloransulam, and imazamox which are members of the sulfonylurea, triazolopyrimidine, and imidazolinone chemical families, respectively. A log scale of eight rates was applied to common ragweed biotypes centering on the current labeled rate for each product. Log-logistic analysis was used to identify LD_{50} values and derive a level of resistance ratio. A lethal dose of glyphosate was unable to be attained for the resistant population whereas the susceptible populations from Edgecombe County and Mississippi experienced 50% mortality with 1.88 and 3.75 kg ae, respectively. The resistant population was also found to be 26 to 71-fold more resistant than the susceptible populations when treated with three families of ALS- and PPO-inhibiting herbicides.

Results from this study indicate the multiple-resistant common ragweed biotype displays cross-resistance to three families of ALS-inhibitors and three families of PPO-inhibitors, further reducing the number of viable control options available to North Carolina farmers.

SWEET POTATO TOLERANCE AND WEED CONTROL WITH EXPERIMENTAL HERBICIDES.
B. Scott* and M. VanGessel, University of Delaware, Georgetown, DE (18)

ABSTRACT

Sweet potato is a full season crop which requires early season weed control. Few options are available which provide effective weed control with minimal crop injury. This trial was established to evaluate three potential herbicides for use in sweet potato.

The experiment was conducted for 3 years at the University of Delaware's Research and Education Center (2015 to 2017). Experiment design was a randomized complete block with ten treatments and three replications. Field was chisel plowed and disked prior to transplanting. Plots were two rows (5-ft. apart), 25 ft. long. Sweet potato slips, 'Beauregard', were transplanted in early June at 1 plant/ft. of row. Sweet potatoes were hilled once and cultivated 1-2 times each year at 3-4 WAP. Layby nitrogen (25 lbs/A) was applied prior to cultivation. Treatments consisted of Linex, Reflex, Valor SX, and Dual Magnum the first year. In subsequent years fluridone was added to the trial. Herbicide rates and timings listed in Table 1. No treatment included an adjuvant.

Table 1. Trial treatments.

Herbicide	Rate/A	Timing
Check		
Fluridone low rate	0.15 or 0.2 lbs ai	PRE-transplanting
Fluridone high rate	0.3 or 0.4 lbs ai	PRE-transplanting
Linex	16 fl oz	28 days after transplanting
Linex	32 fl oz	28 days after transplanting
Linex fb Linex	16 fb 16 fl oz	14 fb 28 days after transplanting
Linex fb Linex	32 fb 32 fl oz	14 fb 28 days after transplanting
Reflex	1 pt	PRE-transplanting
Reflex	2 pt	PRE-transplanting
Valor SX fb Dual Magnum	2.5 oz fb 1 pt	PRE-transplanting fb 1 day after transplanting

Weed control with fluridone was not as effective as the standard treatment (Valor SX fb Dual Magnum) but minimal difference in control did not result in yield loss. Linex as a single application resulted in poor weed control. Linex sequential provided weed control comparable to the standard treatment except for late-season grasses. Reflex provided weed control comparable to the standard treatment except was less effective on morningglory species in 2015 and smooth pigweed in 2017.

Significant early season injury occurred with fluridone. In 2017 at 3 WAT the low (0.2 lb ai/A) and high (0.4 lb ai/A) rates of fluridone resulted in 25 and 40% leaf burn, respectively. In 2017 sweet potato plants treated with the high rate of fluridone remained stunted (>25%) at 7 WAT. In 2016 at 3 WAT the high rate of fluridone resulted in significant leaf burn (16%). Although not significantly different from the standard (Valor SX fb Dual Magnum), greater than 30% stunting was observed with the high (0.3 lb ai/A) rate of fluridone at 4 WAT in 2016. In 2015 and 2016 injury with Linex was comparable to or slightly less injurious than the standard with the exception of the 32 fl oz sequential treatment which resulted in 18% leaf burn at 3 weeks after transplanting (1 WAT). In 2017 Linex at the 32 fl oz rate resulted in a significant amount of stunting which still remained visible at 7 weeks after transplanting (3 and 5 WAT). Reflex injury was comparable to or less than the standard treatment for all three years.
No yield reduction resulted from either rate of fluridone in 2016. However in 2017 the high rate of fluridone resulted in a 24% yield loss. In 2015 and 2016 Linex treatment yields were comparable to the standard treatment. However, in 2017 only the Linex at the 16 fl oz sequential rate produced yields comparable to the standard. All other Linex treatments resulted in reduced yields in 2017. Yield with Reflex regardless of rate did not differ from the standard either year.

Early season injury with fluridone is a concern. Linex sequential treatment at the 16 fl oz rate and Reflex showed potential as herbicide options for sweet potato.

PALMER AMARANTH RESPONSE TO PREEMERGENCE HERBICIDES IN NORTH CAROLINA.
J.W. Vetter*, W.J. Everman, B. Schrage, and J. Sanders, North Carolina State University, Raleigh, NC (20)

ABSTRACT

The overreliance on glyphosate, PPO- and ALS-inhibiting herbicides as POST control in soybeans has contributed to suspected multiple and cross resistant Palmer amaranth populations in North Carolina. With POST control options becoming increasingly limited, producers will need to increasingly rely on PRE herbicides to effectively control herbicide-resistant populations of Palmer amaranth and reduce the selection of resistance to additional modes of action. Metribuzin was once a widely used PRE herbicide in soybeans but variability associated with varietal response and release of effective POST options have limited its use in North Carolina. The need for alternative and multiple modes of action to combat herbicide resistance has gained new prominence and therefore led to a renewed interest in using metribuzin as a PRE herbicide in soybean production.

In 2017, a field study was conducted to evaluate effective metribuzin application rates on soybean tolerance and control of a suspected glyphosate, ALS- and PPO-resistant Palmer amaranth population. Twelve rates, ranging from 0 to 4480 g ai ha^{-1} were applied immediately after planting. At 4 WAT, metribuzin rates above 560 g ai ha^{-1} provided greater than 60% control of Palmer amaranth. Rates above 1120 g ai h^{-1} provided greater than 90% control at 4 WAT. At 6 WAT, rates above 1120 g ai ha^{-1} provided greater than 75% control with greater than 90% control at rates above 2240 g ai ha^{-1}. Soybean injury increased from 10 to 32% when rates increased from 1120 to 4480 g ai ha^{-1} but by 6 WAT no visual injury symptoms were detected in soybeans.

Results from these studies indicate metribuzin may be a viable option at reduced rates in sandy soils of the Coastal Plain of North Carolina to provide both control of Palmer amaranth and crop safety to soybean.

EVALUATION OF POSTEMERGENCE HERBICIDES FOR CONTROLLING CAROLINA REDROOT (*LACHNANTES CAROLIANA*) IN CRANBERRY (*VACCINIUM MACROCARPON*). B. Carr*, P.V. Oudemans, and T.E. Besançon, Rutgers, Chatsworth, NJ (21)

ABSTRACT

29.6 million kg of cranberries in 2016 at a farm value of $28 million (USDA 2017). Cranberry beds in New Jersey are concentrated in the Pine Barrens coastal plain where soil conditions (sandy texture, pH 4.0 to 5.0, good drainage) are optimal for cranberry production. The perennial nature of cranberry production predisposes the crop to a diversity of weed species ranging from herbaceous weeds to woody perennial species. Among perennial weed species, Carolina redroot has been an increasing source of concern for New Jersey cranberry growers regarding the lack of sufficient control from their current management strategies. Carolina redroot is a perennial herbaceous monocotyledonous species member of the *Haemodoraceae* family whose common name is derived from the orange to red coloration of its roots and rhizome. Information regarding herbicidal control of Carolina redroot is extremely limited and mostly restricted to blueberry production (Myers et al. 2013). In order to address the issues of successfully managing Carolina redroot under extremely specific environmental and cropping conditions, a study was initiated in 2017 for evaluating the efficiency of ten postemergence herbicides to control Carolina redroot. Diquat at 560 g ai ha^{-1} provided over 90% control up to 21 days after treatment (DAT) but declined after 42 DAT with the emergence of new shoots. Control with mesotrione at 280 or 560 g ai ha^{-1} increase from less than 15% 14 DAT to 97% 63 DAT. Control increased from 14 to 63 DAT with flumioxazin at 210 g ai ha^{-1} (73%) and 2,4-D at 1,280 g ae ha^{-1} (60%) whereas glyphosate at 1,260 g ae ha^{-1} did not achieve more than 48% control 63 DAT. Other herbicides tested (pronamide at 2,240 g ai ha^{-1}, clopyralid at 70 or 140 g ai ha^{-1}, quinclorac at 1,280 g ae ha^{-1}, carfentrazone at 35 g ai ha^{-1}, and fomesafen at 420 g ai ha^{-1}) provided less than 30% control from 21 DAT onwards. Control of Carolina redroot rhizomes/roots was greatest in plants treated with mesotrione (> 95%), glyphosate (90%), diquat (89%), 2,4-D (84%), and flumioxazin (78%). Nontreated Carolina redroot shoot and root/rhizome fresh weight were 5.8 and 7.7 g, respectively. Greatest shoot and root/rhizome biomass reductions were noted with diquat, mesotrione at 280 or 560 g ai ha^{-1}, glyphosate, 2,4-D, and flumioxazin (0.7 and 0.7 g, 0.4 and 1.8 g, 0.3 and 1.2 g, 0.8 and 1.1 g, 1.1 and 1.3 g, 1.5 and 1.5 g respectively). Clopyralid at 70 or 140 g ai ha^{-1} and fomesafen also reduced root/rhizome fresh weight (4.8, 5.7, and 3.1 g, respectively) but did not decrease shoot biomass. No secondary shoots growing from the rhizome were noted for plants treated with glyphosate, mesotrione, 2,4-D and flumioxazin.

WEED MANAGEMENT IN GLUFOSINATE-RESISTANT SOYBEANS: IT'S ALL ABOUT TIMING.
M. VanGessel*, Q. Johnson, and B. Scott, University of Delaware, Georgetown, DE (23)

ABSTRACT

[no abstract submitted]

COMPARING WEED COMMUNITIES IN ANNUAL AND PERENNIAL SMALL GRAIN CROPPING SYSTEMS DURING THE ESTABLISHMENT YEAR. E. Law*, A. DiTommaso, M. Ryan, S. Wayman, and C. Pelzer, Cornell University, Ithaca, NY (24)

ABSTRACT

Perennial varieties of small grain crops are a nascent but promising development for long-term sustainability of global food production. Small grain crops comprise more than 70% of all agricultural production and are the primary source of calories and nutrition for the majority of humanity. By utilizing perennial analogs of grain species such as wheat and rye, small grain cropping systems may be able to reduce fuel usage and labor while also protecting soil via reduced tillage and continuous ground cover. Optimizing cropping systems designed for perennial grain production is a critical step in increasing their adoption and thereby realizing potential environmental benefits, with management of weed competition being a major issue to be addressed. Here we report on differences in species richness and total biomass observed in weed communities between two perennial and two annual small grain crops grown in a field experiment. Weed communities appear to be more strongly affected by particular crop species and establishment success than perenniality during the first year of the experiment. Intercropping red clover also appeared to enhance weed suppression for all grain crops without negatively impacting yields. In the future we hope to use these data to inform development of integrated weed management programs for both conventional and organic perennial grain programs in the Northeast US.

DOES AMINOCYCLOPYRACHLOR PROVIDE EFFECTIVE CONTROL OF MILE-A-MINUTE (*PERSICARIA PERFOLIATA*)? A. Causey*, A.E. Gover, and M.L. Fuchs, Penn State, State College, PA (29)

ABSTRACT

Mile-a-minute weed (*Persicaria perfoliata (L.) H. Gross)* is a thorny, herbaceous, vining annual native to Asia. It is fast growing and can form dense mats over existing vegetation. Aminocyclopyrachlor at 0.05, 0.11, 0.16, or 0.21 kg/ha was compared to aminopyralid at 0.06 kg/ha, triclopyr at 0.84 or 1.1 kg/ha, glyphosate at 1.1 kg/ha, and glufosinate at 1.03 kg/ha in a floodplain at Bald Eagle State Park in Howard, PA. Methylated seed oil surfactant was added at 1 percent v/v. Treatments were applied at 187 L/ha on July 7, 2017 to four replicates of 2.7 m by 4.6 m plots using a randomized complete block design. At the time of application, the vegetation averaged 1.4 m tall and was comprised mostly of Japanese stiltgrass (*Microstgium vimmineum* (Trin.) A. Campus*)*, wildrye species (*Elymus* spp L.), goldenrod species (*Solidago* spp L.), rice cutgrass (*Leersia oryzoides* (L.) Sw.), and beggarslice (*Hackelia virginiana* (L.) I.M. Johnston). The overstory was scattered black walnut (*Juglans nigra* L.) and black cherry (*Prunus serotina* Ehrh.).

The plots were visually evaluated 0, 31, and 63 days after treatment (DAT) for percent total cover and mile-a-minute weed cover. Mile-a-minute weed injury was recorded 31 DAT. All plots started with similar amounts of mile-a-minute weed cover and total cover ranging from 7 to 10 percent and 96 to 98 percent respectively. The percent of mile-a-minute weed cover averaged 29 percent in the untreated plots 63 DAT. The percent mile-a-minute cover for aminocyclopyrachlor at 0.05 kg/ha (15 percent) was not significantly less than the control, but the percent cover for all other treatments (0 to 6 percent) was significantly less than the control. This suggests that aminocyclopyrachlor applied at rates of 0.11 kg/ha or higher does not differ from standards in its control of mile-a-minute weed. The injury to mile-a-minute weed was rated on a 1 to 10 scale (1 represents no injury and 10 represents complete mortality). All treatments except triclopyr at 0.84 kg/ha caused significantly more injury than the control. Injury to mile-a-minute weed increased as the rates of aminocyclopyrachlor increased. The percent total cover for the untreated plots averaged 94 percent 63 DAT. The percent total cover for the aminocyclopyrachlor at 0.05 kg/ha, aminopyralid, glufosinate, and both triclopyr treatments did not significantly differ from the control (80 to 97 percent). All other treatments had significantly less total cover than the control (30 to 65 percent).The primary species and the next three dominant species were recorded for each plot. On average, Japanese stiltgrass was the most prevalent species for each treatment 0 DAT. Japanese stiltgrass decreased in the aminocyclopyrachlor at 0.05, 0.16, or 0.21 kg/ha and glyphosate treatments 63 DAT, but it increased in the aminopyralid and triclopyr at 0.84 kg/ha treatments. When treating mile-a-minute in areas with Japanese stiltgrass, aminocyclopyrachlor suppresses both species while retaining safety to most other grasses and graminoids, providing an efficacious tool that may provide unique advantages in certain species assemblages.

HERBICIDE-RESISTANT ITALIAN RYEGRASS MANAGEMENT IN VIRGINIA. C.R. Ervin*, C.W. Cahoon, M.S. Reiter, and T. Hines, Virginia Tech, Painter, VA (30)

ABSTRACT

Italian ryegrass (*Lolium perenne ssp. multiflorum)* is one of the most prevalent and troublesome weeds infesting winter wheat. Traditionally, ACCase-inhibiting (diclofop and pinoxaden) and ALS-inhibiting (mesosulfuron and pyroxsulam) herbicides have controlled Italian ryegrass well. In recent years, Virginia growers have claimed Italian ryegrass control by these herbicides to be poor. Currently, only diclofop-resistant Italian ryegrass has been confirmed in Virginia. The objective of this research was to determine the distribution of herbicide-resistant Italian ryegrass in eastern Virginia. During summer 2016, Italian ryegrass seed were collected from 82 winter wheat fields in eastern Virginia (east of I-95) including a known susceptible biotype from Painter, VA. Approximately 400 and 100 to 150 seeds from each sample were planted in greenhouse plots during fall 2016 and fall 2017, respectively. Italian ryegrass was allowed to grow 3.5 to 4 inches tall (1 to 2 leaf stage) prior to treating with herbicides. Herbicide treatments included diclofop-methyl at 1 lb ai/A, mesosulfuron at 0.013 lb ai/A, pinoxaden at 0.54 lb ai/A, pyroxsulam at 0.016 lb ai/A, and no herbicide. Treatments were replicated twice. Visual estimates of Italian ryegrass control were collected 28 and 42 to 56 days after treatment (DAT) and Italian ryegrass fresh biomass was collected 42 to 56 DAT. During 2017, Italian ryegrass density in each pot was collected prior to herbicide treatment and again 56 DAT to determine percent survival. Of the 82 samples, 23 and 30% were resistant to pinoxaden and diclofop-methyl, respectively. Most samples resistant to diclofop-methyl were also resistant to pinoxden, however, for 6 samples, pinoxden remained effective despite poor diclofop-methyl activity. ALS-resistant Italian ryegrass was more widespread throughout eastern Virginia. Of the surveyed populations, 92 and 93% were resistant to mesosulfuron and pyroxasulam, respectively. Several Italian ryegrass biotypes were resistant to all four herbicides. For control of these biotypes, pyroxasulfone applied delayed-preemergence would be in order.

COMPARISON OF DICAMBA- VERSUS GLUFOSINATE-BASED HERBICIDE PROGRAMS IN SOYBEANS. L.S. Rector*, M. Flessner, and K. Pittman, Virginia Tech, Blacksburg, VA (31)

ABSTRACT

Increasing cases of weeds with glyphosate resistance has made weed management more difficult in glyphosate-based systems, prompting the need for alternatives. Postemergent (POST) weed control in soybeans is important to protect yields and reduce weed populations for future years. Research was conducted to assess and compare weed control of dicamba- and glufosinate-based herbicide programs in soybeans.

The study was conducted in Blacksburg, Virginia, and implemented as a randomized complete block design with four replications. The study consisted of eight treatments including two nontreated checks: one for the dicamba resistant system (Roundup Ready 2 Xtend) and one for the glufosinate resistant system (Liberty Link). Treatments consisted of a preemergent (PRE) of flumioxazin (Rowel) at 0.07 kg ha^{-1} followed by (fb) dicamba (Xtendimax with VaporGrip) at 0.56 kg ae ha^{-1} + Intact at 0.5% v v^{-1}, + glyphosate at 1.27 kg ae ha^{-1} (Roundup Powermax), a PRE of flumioxazin + dicamba fb dicamba + Intact + and glyphosate, a PRE of flumioxazin + dicamba fb dicamba + Intact + glyphosate + acetochlor (Warrant) at 1.26 kg ha^{-1}, a PRE of flumioxazin fb glufosinate (Liberty) at 0.60 kg ha^{-1}, a PRE of sulfentrazone at 0.28 kg ha^{-1} + chlorimuron ethyl at 0.018 kg ha^{-1} (Authority Maxx) fb glufosinate, and a PRE of sulfentrazone + chlorimuron ethyl fb glufosinate + pyroxasulfone (Zidua) at 0.12 kg ha^{-1}. Herbicide products and rates were constant across all treatments. Soybeans were planted on 76 cm rows in a no-tillage system for each respective herbicide program. PRE treatments were applied on May 31st, 2017, directly after planting, and POST treatments were applied on June 22nd when weeds were ~7.6 cm tall. Plots were 4 rows wide by 7.6 meters. Treatments were applied with a hand boom at 112 L ha^{-1}. Visible crop injury and weed control were evaluated on a 0 (no control) to 100% (complete necrosis) scale 14 and 39 days after POST application (DAP). Weeds rated were pitted morningglory *(Ipomoea lacunosa)*, redroot pigweed *(Amaranthus retroflexus)*, large crabgrass *(Digitaria sanguinalis)*, and giant foxtail *(Setaria faberi)*. Giant foxtail was only rated 39 DAP. Data were subjected to ANOVA followed by means separation using Fishers Protected $LSD_{(0.05)}$. Data from nontreated plots were excluded from the analysis.

Both programs provided >75% pitted morningglory control and >95% redroot pigweed control 39 DAP with no differences detected. Within the glufosinate-based system, POST glufosinate treatments with a grass controlling residual herbicide increased control of annual grasses by >10% when compared to glufosinate alone. Across grass species, no control differences were observed between dicamba- and glufosinate-based systems. Off-target dicamba movement resulted in 7 to 15% damage to glufosinate resistant soybeans 39 DAP. To increase control of grass species, a residual herbicide should be added to glufosinate-based herbicide programs. This research indicates that both dicamba- and glufosinate-based systems resulted in similar, effective weed control. Future research is needed to corroborate these findings across multiple site-years and determine control in glyphosate resistant weed populations.

EFFICACY OF AMINOCYCLOPYRACHLOR FOLIAR TREATMENTS AGAINST SEVERAL EXOTIC SHRUB SPECIES. M.L. Fuchs, A.E. Gover*, and A. Causey, Penn State, University Park, PA (34)

ABSTRACT

Aminocyclopyrachlor (ACP) is a synthetic auxin first made available for non-crop use in 2011. ACP is labeled for terrestrial applications commonly used in utility rights-of-way, where grass-safe, persistent control of woody vegetation is highly desired. ACP is not currently a staple product in natural resource management in the northeast U.S. Factors in its favor include low toxicity, flexible application (water's edge), and despite being generally safe to graminoids, it is quite active against Japanese stiltgrass (*Microstegium vimineum* (Trin. A. Camus), a common exotic pest in natural areas. Factors against adoption include significant soil activity and persistence, and lack of data regarding key target susceptibility. Separate individual plant treatment experiments were conducted against common buckthorn (*Rhamnus cathartica* L., RHACT), autumn olive (*Elaeagnus umbellata* Thunb., ELGUM), and Japanese barberry (*Berberis thunbergii* DC., BEBTH). Application was low-volume (235 L/ha), using a CO_2-powered, hand-held sprayer with an adjustable cone nozzle, targeting nine stems per treatment, with stems assigned in a Completely Random Design. ACP was evaluated alone at 0.14, 0.21, 0.28, and 0.56 (BEBTH only) kg/ha; and tank-mixed with imazapyr (0.28 plus 0.21 kg/ha), triclopyr (0.28 (ACP) plus 0.21, or 0.42 (ACP) plus 0.21 kg/ha), triclopyr + fosamine (028 (ACP) plus 0.21 plus 3.4 kg/ha, respectively), and as the premixes ACP plus metsulfuron (0.26 plus 0.084 kg/ha, respectively) and ACP plus metsulfuron plus imazapyr (0.26 plus 0.084 plus 0.35 kg/ha, respectively). The RHACT and ELGUM experiments included an operational standard (glyphosate plus triclopyr plus imazapyr at 3.4 plus 1.3 plus 0.14 kg/ha, respectively). The BEBTH study included standalone treatments of aminopyralid (0.19 kg/ha), triclopyr (2.5 kg/ha), and glyphosate (3.4 kg/ha). RHACT was treated at the Penn State Arboretum, University Park, PA, August 12, 2015; ELGUM was treated at Canoe Creek State Park, Hollidaysburg, PA, on August 13, 2015; and BEBTH was treated at Bald Eagle State Park, Howard, PA, on July 10, 2017. Response was visually rated as defoliation and mortality, approximately 2 months after treatment (MAT), and approximately 12 and 24 MAT. Data were subjected analysis of variance, and means separated using Fisher's Protected LSD if treatment effect was significant.

For RHACT, the operational standard was rated at 72 percent defoliation and 29 percent mortality 24 MAT. The best-rated ACP treatments (99 percent defoliation, 88 percent mortality) were applied at 0.26 to 28 kg/ha, and were rated significantly higher than the standard for both metrics. There were no significant differences between any ACP-containing treatments. For ELGUM, the standard provided 100 percent mortality at 12 MAT. The best-rated ACP treatments provided 88 to 100 percent mortality. Treatment effect was non-significant for both defoliation and mortality (untreated checks excluded from analysis). In the first-season evaluation for BEBTH, there was an ACP rate effect, with 0.28 and 0.56 kg/ha rated significantly higher (97 and 100 percent, respectively) for defoliation than 0.14 kg/ha (77 percent). The aminopyralid, triclopyr, and glyphosate treatments ranged from 89 to 100 percent defoliation, and were not significantly different from the best-rated ACP treatments.

In a spot-treatment scenario with the most common exotic shrubs in PA State Parks, ACP does not offer advantage over the current standard, which relies on aquatic-labeled, low-persistence herbicides. It will have utility where selectivity to a grass groundcover is required, and where concurrent, selective targeting of Japanese stiltgrass is an objective.

EVALUATING PREEMERGENCE HERBICIDES IN PUMPKINS. J.H. Ferebee IV*, C.W. Cahoon, T. Hines, M. Askew, and H.B. Blake, Virginia Tech Graduate Research Assistant (Dr. Cahoon), Painter, VA (35)

ABSTRACT

Direct-seeded pumpkin production is popular in Virginia despite weed control being more difficult than when the crop is produced in a plasticulture system. Residual herbicides are the foundation of weed control in direct-seeded pumpkin. To expand residual options, the emulsifiable concentrate (EC) and micro-encapsulated (ME) formulations of acetochlor and fluridone were compared to common preemergence herbicides for pumpkin tolerance. Pumpkin cultivar "Kratos" was planted into conventionally prepared land on June 13, 2017 near Painter, VA and July 6, 2017 near Virginia Beach, VA. Preemergence treatments were applied immediately after planting using a CO_2-pressurized backpack sprayer. Treatments included fomesafen at 0.19, fomesafen at 0.25, ethalfluralin at 0.56, clomazone at 0.26, halosulfuron at 0.035, fluridone at 0.15, *S*-metolachlor at 0.95, EC acetochlor at 1.125, and ME acetochlor at 1.125 lb ai/A. A non-treated check was included for comparison. Pumpkin stand was collected 14 days after planting (DAP). Visual estimates of pumpkin tolerance and weed control were collected 14, 28, 42, and 56 DAP. Pumpkins were harvested and weighed at the end of the season to determine total fruit number, average fruit size, and total yield. Fomesafen and halosulfuron reduced pumpkin stand approximately 20% compared to the non-treated check at Painter whereas pumpkin stand was similar among all treatments at Virginia Beach. Pumpkin injury was greatest 14 DAP. At this timing in Painter, fluridone, EC acetochlor, and ME acetochlor injured pumpkin 44, 31, and 33%, respectively. All other treatments at this location injured pumpkin 20% or less. Fourteen DAP at Virginia Beach, fluridone injured pumpkin 26%, ME acetochlor 32%, and EC acetochlor 53%. Fluridone and both formulations of acetochlor were much less injurious later in the growing season. Despite early season injury, fluridone controlled ivyleaf morningglory 85% at Painter and pitted morningglory 95% at Virginia Beach 28 DAP. All other treatments were less effective controlling morningglory species. Similarly, at Painter 28 DAP, fluridone controlled spurred anoda 85% and was more effective than all other treatments except clomazone (88%). Average fruit size, total fruit number, and total yield ranged 20 to 22 lb/fruit, 2359 to 2613 fruit/A, and 48521 to 54450 lb/A, respectively, for all treatments except *s*-metolachlor, EC acetochlor, ME acetochlor, and the non-treated check. For these four treatments, average fruit size ranged 15 to 18 lb/fruit, total fruit number 1996 to 2323, and total yield 31944 to 39688 lb/A. Although fluridone injured pumpkin early in the season, plots treated with the herbicide yielded similarly to plots treated with commercial standard residual herbicides. In addition, fluridone controlled morningglory species and spurred anoda well. Preliminary results suggest fluridone may have utility for pumpkin production. However, additional research is needed to determine pumpkin tolerance to various rates of fluridone and under varying environmental conditions and soil types.

FROM THE EDGE AND BEYOND: PERFORMANCE OF JOHNSONGRASS AT THE BOUNDARY OF ITS DISTRIBUTION. R.A. Fletcher*, D.Z. Atwater, D. Haak, and J. Barney, Virginia Tech, Blacksburg, VA (36)

ABSTRACT

Phenotypic plasticity and local adaptation play an important role in the range expansion of invasive plants and agricultural weeds. The ability of a plant to adapt to novel environmental conditions allows it to spread and invade new areas. However, all species have limits to their geographic distributions and may experience allocation trade-offs beyond their range boundary (e.g., a trade-off between growth and reproduction). Johnsongrass (*Sorghum halepense* (L.) Pers.) is a globally invasive weed. Studies have demonstrated significant phenotypic variation among populations throughout its North American distribution. In the United States, Johnsongrass is most abundant and problematic in the eastern half of the country below 38°N. Above 38°N, abundance decreases precipitously. We had sought to determine if there is variation in performance and allocation trade-offs among Johnsongrass populations collected from populations throughout its US distribution and grown in a common location at the range boundary. We chose populations from the distribution core: Georgia, where Johnsongrass is most abundant and problematic; the boundary: Virginia and Texas, beyond which abundance decreases precipitously; and the periphery: New York and New Mexico, where populations are small and extremely rare. In May 2017, we established a common garden experiment in Virginia, which is near the range boundary, using the five collected populations replicated 10 times. On a weekly basis during the growing season, we recorded days to flowering and mortality, and in October 2017, we measured height, counted the number of flowering culms, and harvested, dried, and weighed the aboveground biomass. Preliminary results indicate that performance differed among the populations with the Virginia and Texas populations growing much larger than the New York, New Mexico, and Georgia populations. We also found the New York population invested more in reproduction than the other populations. These results suggest there is phenotypic variation in performance among the populations of Johnsongrass and provide some evidence consistent with local adaptation. Additionally, there may be a trade-off between growth and reproduction for the New York population when grown at the range boundary.

MECHANICAL AND CHEMICAL PROGRAMS FOR BERMUDAGRASS CONTROL IN CREEPING BENTGRASS. J. Craft*, J. Brewer, and S. Askew, Virginia Tech, Blacksburg, VA (37)

ABSTRACT

Bermudagrass (*Cynodon dactylon* L) is a persistent and troublesome weed in bentgrass fairways throughout the transition zone. The persistence of bermudagrass in bentgrass fairways reduces turf quality due to differences in growth habit, leaf texture, and color. Selectively controlling bermudagrass in bentgrass is challenging because most herbicides that control bermudagrass are often injurious as well. Previous research suggest proper timing can maximize bermudagrass control in the fall when creeping bentgrass is actively growing and bermudagrass is going into dormancy. However, no research has examined a combination of mechanical and chemical programs for bermudagrass control in the fall. Therefore, the objective of this research was to evaluate herbicide and mechanical slicing programs for bermudagrass control in bentgrass fairways. Three field trials were initiated in the fall of 2017 and will continue into the spring of 2018 to evaluate herbicide and slicing programs for bentgrass tolerance and bermudagrass control. Three 8 treatment trials were initiated on September 5, 2017 with two being at the Glade Road Research Facility in Blacksburg, VA and one at the Ballyhack Golf Course in Roanoke, VA. In Blacksburg, VA, one trial was established on a pure stand of creeping bentgrass fairway to evaluate tolerance with the other trial being established on a pure stand of Patriot bermudagrass fairway to evaluate bermudagrass control. The trial at Ballyhack Golf Course in Roanoke, VA was established on a creeping bentgrass fairway that was infested (40 to 50%) with bermudagrass to evaluate both bentgrass tolerance and bermudagrass control. Treatments were arranged as a split plot design with four replicates. All treatments consisted of three applications at two week-intervals accompanied by slicing. Three additional slicing occurred 2, 4, and 8 weeks after the last herbicide treatment. Treatments programs included four main plots: non-treated, topramezone at 6.4 g ai ha^{-1} + triclopyr at 26 g ai ha^{-1}, ethofumesate at 841 g ai ha^{-1}, topramezone at 6.4 g ai ha^{-1} + triclopyr at 26 g ai ha^{-1} + ethofumesate at 841 g ai ha^{-1}, and two subplots, which are slicing and no slicing. All topramezone-containing treatments included a methylated seed oil adjuvant at 0.5% v v^{-1}.

At 8 weeks after initial treatment (WAIT), topramezone + triclopyr with or without slicing controlled bermudagrass 80%. Ethofumesate alone was the slowest acting treatment initially. However, at 8 WAIT, control of bermudagrass was similar to topramezone + triclopyr. The addition of slicing to ethofumesate treatments appeared to increase bermudagrass control. The addition of ethofumesate to topramezone + triclopyr increased control of bermudagrass greater than 90%. Topramezone + triclopyr + ethofumesate was the most effective treatment at controlling bermudagrass. Slicing increased control initially but by 8 WAIT there was no difference observed between slicing and no slicing. The topramezone + triclopyr + ethofumesate treatment was also the most injurious treatment causing 40% injury with the addition of slicing causing more injury than no slicing. The evaluations following spring green-up will more clearly indicate which herbicide programs were the most effective and if the addition of slicing increases bermudagrass control.

MICROSTEGIUM VIMINEUM POPULATION RESPONSE AND ECONOMICS FOLLOWING SEVEN YEARS OF MANAGEMENT. J. Brewer*, B. Fletcher, J. Barney, S. Askew, D. Tekiela, and A. Post, Virginia Tech, Blacksburg, VA (38)

ABSTRACT

Japanese stiltgrass (*Microstegium vimineum*) is a nonnative, invasive grassy weed that has spread up and down the eastern coast of the United States from New York to Florida and as far west as Texas. It is considered one of the worst invaders by many agencies at the local, state, and federal levels. *Microstegium vimineum* is detrimental to forest ecosystems as it outcompetes native flora and alter soil characteristics which cause reductions in biodiversity. There are many studies that indicate that both chemical and nonchemical programs can effectively control *M. vimineum* on a year to year basis. Unfortunately, these programs do not have long-term effect on the *M. vimineum* population due to the longevity of seed viability in the soil seedbank. Researchers have observed its seed persisting for up to three years in the soil. At this time, there are no long-term studies evaluating how many years of continual management is required to effectively eradicate *M. vimineum* from a targeted area or the economic burden of continuous management. At Virginia Tech, we have an ongoing seven year study evaluating five different management programs for long-term control of *M. vimineum* and the overall economic cost including labor, fuel, and chemical. The study was initiated in 2011 at a forest site in Newport, VA. The study was set up in a split-block design with four blocked replications, each plot was 42.5 m by 7.6 m. The chemical treatments were applied using a pump-up backpack sprayer that utilized a special pressure regulator, which calibrated the sprayer to apply 140 L/ha. The treatments included pendimethalin at 2.24 kg ai/ha, glyphosate at 2.77 kg ai/ha, low-rate glyphosate at 0.11 kg ai/ha, sethoxydim at 0.772 kg ai/ha, mechanical removal by weed eater, and an untreated check. Each plot was split into upper and lower sections. The upper section of each treatment required two applications before flowering while the lower section was only treated with a single application. Pre-application and post-application *M. vimineum* data was collected via point intercept transects (2 transects = 200 points per plot). Also, two soil samples were taken at random from both the upper and lower sections of a plot to evaluate *M. vimineum* population change in the soil seedbank over time. For economic assessment, there were three main data types collected, which included the amount of timer per application per plot, milliliters of product used per plot, and milliliters of gasoline used during each mechanical removal plot. These data will be combined to calculate an overall price per year per management program.

From 2011 to 2017, we have observed an approximate 80% or greater decrease in *M. vimineum* population from all chemical programs and 60 to 70% decrease in population by mechanical removal. We also are observing similar trends with the seedbank data. In 2016, the two application programs of all treatments reduced *M. vimineum* populations by approximately 70% when compared to the untreated check, while the single application program of high-rate glyphosate was not significantly lower than the untreated check. The single application program of low-rate glyphosate, pendimethalin, and sethoxydim had greater than 70% reduction of *M. vimineum* seedbank populations, while mechanical removal only reduced the population approximately 40%. These data are potentially indicating that the second application is adding a significant benefit to population reduction for certain programs. Even though we have significantly reduced the *M. vimineum* population, these data show that our programs have still not effectively eradicated it from the trial area, which could be due to reintroduction from outside sources via animals or environmental factors.

FALSE-GREEN KYLLINGA (*KYLLINGA GRACILLIMA*) CONTROL WITH VARIOUS HERBICIDES IN THE FIELD AND GREENHOUSE. K.H. Diehl*, M.T. Elmore, A.J. Patton, D.P. Tuck, J. Carleo, and J. Sawyer, Rutgers University, New Brunswick, NJ (39)

ABSTRACT

False-green kyllinga (*Kyllinga gracillima*) is a perennial sedge and pervasive weed of cool-season turfgrass. Two greenhouse experiments and four field experiments were conducted to test the efficacy of various post-emergence herbicides against false-green kyllinga in 2017.

Field treatments included both single and sequential applications of: imazosulfuron (420 g/ha), imazosulfuron (740 g/ha), halosulfuron-methyl (70 g/ha), sulfentrazone (110 g/ha), a single application of sulfentrazone (110 g/ha) + imazosulfuron (420 g/ha), and a non-treated control. False-green kyllinga injury was evaluated visually from 0-100% (with 0% being no injury and 100% completely necrotic). 'Control' was determined by assessing the percent cover of false-green kyllinga in each plot compared to the non-treated control at 4, 8, and 12 weeks after initial treatment (WAIT). Grid intersect counts were also conducted at 12 WAIT. For each experiment, data were analyzed in SAS (v9.4) using a single factor RCBD and Fisher's Protected LSD (α=0.05).

Two experiments were conducted in IN, one at Bloomington Country Club (Bloomington, IN) and Victoria National Golf Club (Newburgh, IN) on creeping bentgrass (Agrostis stolonifera) fairways. Applications were made on May 30th and June 30th. Two additional experiments were conducted at Stone Harbor Golf Club in Cape May, NJ. Site 1 was maintained as a rough at 6.35cm and Site 2 maintained as a fairway. Both sites consisted of creeping bentgrass and natural populations of false-green kyllinga. Initial and sequential applications were made on 13 June and 13 July, respectively. Treatments were arranged in a randomized block design and applied using standard small-plot spray equipment at 410 L ha^{-1} in New Jersey and 810 L ha^{-1} in IN.

A greenhouse experiment was conducted to further explore efficacy of single applications of imazosulfuron, halosulfuron-methyl, imazosulfuron and imazosulfuron + carfentrazone-ethyl on false-green kyllinga. Herbicides were applied at 0.1, 0.25, 0.5, 0.75, 1, 2, 4 and 10 times registered use rates for each herbicide. Treatments included: a nontreated control, imazosulfuron (42, 105, 210, 315, 420, 840, 1680 and 4200 g/ha), halosulfuron-methyl (6.8, 17.3 34, 52.5, 70, 140, 280 and 700 g/ha), sulfentrazone (28, 70, 140, 210, 280, 560, 1120 and 2800 g/ha), and sulfentrazone + carfentrazone-ethyl (77, 153, 230, 306, 640, 1230, and 3060 g/ha). Treatments were replicated four times and arranged in a completely randomized design; the experiment was repeated in time. Treatments were applied using a track sprayer at to 12.7 by 12.7cm pots of false-green kyllinga in peat-based growing medium. Visual injury (0-100%) was measured weekly and clipping yield was collected biweekly. Regression analyses were conducted in SAS (v9.4) to characterize the dose response curve for each herbicide.

In NJ field experiments, all treatments containing imazosulfuron provided >97% false-green kyllinga control at 12 WAIT. Two sequential applications of halosulfuron-methyl provided >95% control but the single application only provided 64% control. Sulfentrazone provided 50% in site 2 and no control in site 1 at 12 WAIT. In IN, sequential applications of imazosulfuron reduced kyllinga cover by 100% at 12 WAIT. Sulfentrazone + imazosulfuron reduced kyllinga cover by >89% at 12 WAIT, while sulfentrazone alone reduced cover by 35%. Treatment responses in greenhouse experiments are similar to those observed in field research.

INFLUENCE OF TIMING AND TANK MIXTURES ON CONTROL OF MULTIPLE-RESISTANT COMMON RAGWEED (*AMBROSIA ARTEMISIIFOLIA*). B. Schrage*, W.J. Everman, J. Sanders, and T.N. O'Quinn, North Carolina State University, Raleigh, NC (40)

ABSTRACT

In 2016, a biotype of common ragweed (*Ambrosia artemisiifoia*) in Northeastern North Carolina was confirmed to be resistant to glyphosate, ALS-, and PPO-Inhibiting herbicides. This discovery serves as the first agronomic dicot exhibiting confirmed resistance to three modes of action in North Carolina, and prompted an evaluation into possible management options for growers in Camden, Chowan, Currituck, Gates, Hertford, Pasquotank and Perquimans counties.

Diminishing POST control options are an unfortunate reality with multiple-resistant weeds, therefore an experiment was conducted in Moyock, NC in 2016 and 2017 to evaluate the impact of tank mixture and application timing on control of multiple-resistant common ragweed. A randomized complete block design was implemented with four replications emphasizing a factorial arrangement of treatments with 5 PPO-inhibitors x 5 tank-mix options x 4 common ragweed heights. The five PPO-based herbicides included lactofen, fomesafen, aciflourfen, fomesafen + *S*-metolachlor, or No PPO at standard POST application rates. These PPO-inhibitors were applied in combination with five tank mixtures which included glufosinate, 2,4-D, dicamba, mesotrione, or No Mix and were applied at four common ragweed heights of 5, 10, 15, or 20 cm. Density and biomass data revealed multiple-resistant common ragweed control was numerically improved in all cases when treating smaller individuals. Lactofen was the most effective PPO-based herbicide, although sufficient control was only attained with the inclusion of 2,4-D, dicamba, or glufosinate. Mesotrione did significantly improve common ragweed control, irrespective of tank-mix partner, albeit to a much lesser extent then auxinic and phosphinic acid herbicides. Additionally, mesotrione efficacy was highly correlated with weed size at time of application.

PPO-inhibiting herbicides applied alone are no longer viable options for common ragweed control in areas where multiple-resistant biotypes are found. Adding PPO-inhibiting herbicides to 2,4-D, dicamba, and glufosinate failed improve or reduce control.

INVESTIGATING HERBICIDE USES TO FURTHER INTERSEEDED CEREAL RYE (*SECALE CEREALE)* USE IN VEGETABLE SYSTEMS. E.M. Buck*, D. Robinson, P. Sikkema, R. Van Acker, and N. Soltani, University of Guelph, Ridgetown, ON (41)

ABSTRACT

Vegetable growers express interest in using living mulches but raise concerns regarding crop competition and difficulties managing the living mulch vegetation. Cereal winter rye holds promise as a spring-sown living mulch if two adoption barriers can be removed: weed competition during rye establishment and excessive rye vigor during the growing season. Experiments conducted in Ridgetown, ON in 2016 and 2017 in a snap bean-rye living mulch production system investigated solutions to both problems. The first examined spring-sown rye tolerance to preemergent herbicides used for early season weed control in snap bean. Rye exhibited good tolerance to all tested preemergent herbicides, meaning that viable options exist for curtailing weed pressure during establishment. The second set of trials tracked the response of spring-sown rye to quizalofop-P-ethyl rates. Results from the quizalofop-P-ethyl trials provide guidance for those seeking flexible options to chemically manage rye vigor in a living mulch system.

EFFECTS OF CEREAL RYE BIOMASS ON GERMINATION AND GROWTH OF PALMER AMARANTH *(AMARANTHUS PALMERI)* AND COMMON RAGWEED *(AMBROSIA ARTEMISIIFOLIA)*. L.S. Rector* and M. Flessner, Virginia Tech, Blacksburg, VA (42)

ABSTRACT

Palmer amaranth *(Amaranthus palmeri)* and common ragweed *(Ambrosia artemisiifolia)* are problematic weeds due in large part to herbicide resistance, which necessitates alternative control strategies. Terminated cereal rye cover crop biomass can suppress summer annual weed germination. Research was conducted to evaluate the optimal cereal rye biomass needed to suppress germination and growth of Palmer amaranth and common ragweed.

Studies were conducted in a greenhouse in Blacksburg, Virginia and designed to simulate a rolled cover crop. A randomized complete block design was utilized with four replications per weed species. Treatments were cereal rye biomass levels of 0, 2000, 4000, 6000, 8000, 10000 kg ha^{-1}. Cereal rye biomass was composed of dry stem, leaf, and flower material obtained from mature plants grown in field conditions. Potting mix was used as soil media. All trays (52 by 25cm) had 100 Palmer amaranth or common ragweed seeds, respectively, spread on top of the soil surface to simulate seed shed in a no-tillage cropping system, before placing the cereal rye biomass. Cereal rye biomass was cut to length to fit inside each tray and weighed. Studies were repeated in time and initiated on June 2 and August 24, 2017. Stand counts of weeds were taken weekly for four weeks. Total counts were used for analysis. At week four, all above ground weed biomass was collected, dried, and weighed. Data were analyzed using JMP Pro 13.0. Linear regressions were used to determine relationship between data parameters and cereal rye biomass.

Common ragweed stand counts ranged from 38 to 538 plants m^{-2} across all treatments. Common ragweed germination in the no-cover control was 53% and had an average biomass of 45g m^{-2}. A 59% stand reduction and a 40% biomass reduction of common ragweed was observed at 10000 kg ha^{-1} of cereal rye when compared to the no-cover control. The quadratic equation comparing cereal rye biomass to common ragweed germination is [percent germination = 56.39 – 0.0007*(cereal rye biomass) – $3.058e^{-7}$*(cereal rye biomass)^2] with an r-squared value of 0.59. Palmer amaranth stand counts ranged from 0 to 492 plants m^{-2}. Palmer amaranth no-cover control had a mean germination of 33% and an average biomass of 38g m^{-2}. 10000 kg ha^{-1} of cereal rye cover reduced Palmer amaranth stand by 77% and Palmer amaranth biomass by 44% when compared to the no-cover control. The quadratic equation comparing Palmer amaranth germination to cereal rye biomass is [percent germination = 31.46 – 0.0025*(cereal rye biomass) + $9.487e^{-9}$*(cereal rye biomass)^2] with an r-squared value of 0.45. Overall, cereal rye biomass was effective at reducing germination of common ragweed and Palmer amaranth. Cereal rye cover crop can suppress weed germination, especially in smaller seeded broadleaf weeds. Future research will be conducted in field and greenhouse trials to determine quality and quantity of light penetrating the cereal rye cover crop canopy.

HERBICIDE TOLERANCE OF INDUSTRIAL HEMP (*CANNABIS SATIVA* L.). J. Byrd*, M. Flessner, J. Fike, and K. Bamber, Virginia Tech, Blacksburg, VA (43)

ABSTRACT

Industrial hemp (*Cannabis sativa* L.) is a multipurpose crop that is mainly cultivated for its fiber and seed content. Hemp stalk can be used as a primary source in auto parts, textile items, industrial products, and building materials. Despite these benefits, industrial hemp is only currently being explored as a commercial crop in Virginia. Optimal production practices are not known, including weed control. Therefore, in 2017, greenhouse and field studies were conducted to assess herbicide tolerance of industrial hemp.

Pre-emergence (PRE) and post-emergence (POST) herbicide tolerance greenhouse screenings were conducted separately, and each study was repeated in time. For the PRE screening, 10 seeds of monoecious cultivar 'Felina 32' were sown into 3.78 L pots with 8 replicates for each treatment. For the POST screening, 120 seeds of the same cultivar were planted in cone-tainers (1 seed per cone-tainer). Both studies used a silty loam Ross soil. Plants were watered to maintain field capacity. Data collected included visible plant injury (0 to 100% scale) and above ground biomass yield 8 weeks after treatment.

Field screening trials were established at Virginia Tech's Kentland Farm in early-June in 6.7 m^2 plots (1.8 by 3.7m) and included 5 PRE and 5 POST herbicides, respectively. The PRE screening was in a fiber (27 kg ha^{-1} planting rate with 19 cm row spacing) production setting while POST was in a grain (22 kg ha^{-1} planting rate with 38 cm row spacing). Visible injury and grain yield data were collected. The experimental designs for all studies was a randomized complete block.

Greenhouse data indicate that the PRE herbicides flumioxazin, pendimethalin, metribuzin, and diuron are suitable candidates for further testing having <40% injury. The POST herbicides chlorimuron (40% injury) and halosulfuron (48% injury) are suitable for further testing. The PRE field screening resulted in no differences in grain yields, which ranged from 249 to 364 kg ha^{-1}, despite >50% injury 60 days after treatment from chlorimuron, linuron, and pendimethalin. Other PRE herbicides screened were fomesafen and *S*-metolachlor, which had similar injury to the nontreated. In the POST field study, no differences were detected relative to the nontreated in grain yield, which ranged from 316 to 577 kg ha^{-1}. Halosulfuron was the only POST treatment to result in injury relative to the nontreated (70%, 30 days after treatment). Other POST herbicides screened included sethoxydim, bromoxynil, quizalofop, and clopyralid and had injury similar to the nontreated 21 and 30 days after treatment. Further research is necessary to corroborate these findings across multiple site-years.

TERMINATION OF RAPESEED COVER CROP. M. Askew*, C.W. Cahoon, M. VanGessel, M. Flessner, H.B. Blake, J.H. Ferebee IV, and T. Hines, Virginia Tech, Eure, NC (44)

ABSTRACT

Cover crops have become popular as producers look for alternative ways to suppress weeds and improve soil quality. Commonly used cover crops include cereals, legumes, *Brassicaceae* species, and mixtures of the three. Rapeseed (*Brassica napus* L.) is a popular *Brassicaceae* choice for cover crop mixtures because of its powerful taproot, which creates soil macropores and increases water infiltration. For weed suppression, growers often delay cover crop termination to maximize biomass. Large *Brassicaceae* species can be troublesome to control, and Virginia producers have reported that cover crop rapeseed is difficult to terminate prior to planting soybean. Field experiments were conducted near Painter, VA and Georgetown, DE during 2016-2017 to evaluate various herbicides applied at two timings to simulate early and late burndown scenarios. Dwarf Essex rapeseed was planted on September 26, 2016 in Painter, VA and on October 7, 2016 in Georgetown, DE. Treatments included a factorial arrangement of two burndown timings by 24 herbicide treatments. Early and late burndown timings in Painter, VA were March 20 and April 10, 2017, respectively. In Georgetown, DE, early burndown treatments were applied on April 5 whereas late burndown treatments were applied on April 17, 2017. Herbicide treatments (rate in parenthesis) included 2,4-D low rate (LR) (0.53 kg ae ha^{-1}), 2,4-D high rate (HR) (1.1 kg ae ha^{-1}), dicamba LR (0.28 kg ae ha^{-1}), dicamba HR (0.56 kg ae ha^{-1}) , thifensulfuron + tribenuron premix (0.028 kg ai ha^{-1}), glyphosate (1.27 kg ae ha^{-1}), saflufenacil LR (0.037 kg ai ha^{-1}), saflufenacil HR (0.05 kg ai ha^{-1}), paraquat (0.84 kg ai ha^{-1}), mesotrione (0.105 kg ai ha^{-1}), glufosinate (0.89 kg ai ha^{-1}), glyphosate + 2,4-D LR, glyphosate + dicamba LR, glyphosate + thifensulfuron + tribenuron premix, glyphosate + thifensulfuron + tribenuron premix + 2,4-D LR, glyphosate + thifensulfuron + tribenuron premix + dicamba LR, paraquat + 2,4-D LR, glyphosate + glufosinate, glyphosate + glufosinate + 2,4-D LR, paraquat + mesotrione + atrazine (0.56 kg ai ha^{-1}), paraquat + thifensulfuron + tribenuron premix, glyphosate + saflufenacil HR, glyphosate + saflufenacil HR + 2,4-D LR, glyphosate + saflufenacil HR + dicamba LR. A non-treated check was included for comparison. Visual estimates of rapeseed control were collected 7, 14, and 28 days after each application. Rapeseed biomass was harvested from a 0.25 m^2 28 days after each application and allowed to dry before weighing. Rapeseed in Painter was larger late burndown (92 cm) compared to early burndown (27 cm). Small rapeseed was easier to terminate than larger rapeseed. In general, herbicide treatments were twice as effective when applied early burndown compared to late burndown. Paraquat + atrazine + mesotrione and treatments that included glyphosate + 2,4-D applied early burndown were most effective controlling rapeseed 28 days after early burndown in Painter. However, no treatment controlled rapeseed greater than 80% at this timing. In contrast, no treatment at Painter controlled rapeseed greater than 68% 28 days after late burndown. Trends in rapeseed biomass were similar to visual estimates of rapeseed control. Results observed from Painter during 2017 reinforce the importance of timely rapeseed termination and the use of herbicide combinations, such as glyphosate + 2,4-D, paraquat + 2,4-D, or paraquat + atrazine, to control the species. Control observed in Georgetown was typically higher than control in Painter. This was due to differences in rapeseed size, as rapeseed was 76 cm at late burndown compared to 12 cm at early burndown.

AVOIDING PPO-RESISTANT COMMON RAGWEED IN SOYBEAN. H.B. Blake*, C.W. Cahoon, M. Flessner, M. Askew, J.H. Ferebee IV, and T. Hines, Virginia Tech, Painter, VA (45)

ABSTRACT

Common ragweed (*Ambrosia artemisifolia*) is one of the most troublesome weeds plaguing Virginia soybeans. Glyphosate and acetolactate synthase (ALS) inhibiting herbicides once controlled the weed, however, common ragweed biotypes evolved resistance to these herbicides and now dominate the region. Currently, soybean growers rely on protoporphyrinogen oxidase inhibiting (PPO) herbicides to control common ragweed. Flumioxazin applied preemergence (PRE) followed by fomesafen applied a postemergence (POST) is a popular soybean herbicide program in Virginia. Flumioxazin and fomesafen are both PPO-inhibiting herbicides and growers are concerned over selecting for PPO-resistant common ragweed. Biotypes of common ragweed resistant to PPO-inhibiting herbicides have been confirmed in Delaware, North Carolina, and Ohio. The objective of this project is to evaluate common ragweed control by herbicide programs that included 0, 1, or 2 PPO-inhibiting herbicides with the overarching goal of reducing PPO selection pressure. Studies were conducted near Painter and Suffolk, VA. Soybean cultivar CZ4748LL was planted May 4, 2017 at Painter whereas CZHBK4952LL was planted May 17, 2017 at Suffolk. Treatments consisted of a 6x3 factorial of PRE x POST herbicides. Treatments were replicated 4 times and organized in a randomized complete block design. Preemergence herbicides were applied immediately after planting and consisted of flumioxazin (0.1 lb ai/A), metribuzin (0.23 lb ai/A), linuron (0.22 lb ai/A), acetochlor (1.125 lb ai/A), clomazone (0.47 lb ai/A), and no PRE herbicide. Postemergence treatments were applied approximately 4 weeks after planting and consisted of fomesafen + MSO (0.38 lb ai/A + 1 % V/V), glufosinate + AMS (0.59 lb ai/A + 3 lb/A), and no POST herbicide. Flumioxazin controlled common ragweed 95% just prior to POST. Clomazone (68%), linuron (63%), metribuzin (16%), and acetochlor (19%) were less effective than flumioxazin at the same timing. In the absence of a PRE herbicide, fomesafen controlled common ragweed 100% 4 weeks after POST whereas glufosinate alone POST controlled the weed 88%. However, when using an effective PRE herbicide, common ragweed control by fomesafen and glufosinate differed little. Flumioxazin followed by fomesafen controlled common ragweed 100% 4 weeks after POST, however, this placed increased PPO selection pressure on common ragweed. A more sustainable approach would be to limit PPO inhibitor use to once per growing season. Treatments that included only one PPO herbicide but controlled common ragweed similarly to flumioxazin followed by fomesafen included flumioxazin followed by glufosinate and all treatments that included fomesafen POST. For Roundup Ready soybean, it appears fomesafen POST is the part of the "PPO puzzle" to keep in fields infested with glyphosate- and ALS-resistant common ragweed, while at the same time limiting PPO selection pressure. However, growers can effectively limit PPO selection pressure and still use flumioxazin PRE if they combine it with other residual products (clomazone, metribuzin, linuron) that have common ragweed activity. Alternatively, growers producing Liberty Link soybean and wanting limit PPO selection pressure can effectively control common ragweed with flumioxazin PRE followed by glufosinate POST.

IMPACT OF COVER CROPS ON WEED POPULATIONS AND SOIL HEALTH IN HIGHBUSH BLUEBERRY (*VACCINIUM CORYMBOSUM*). K. Brown*, T.E. Besançon, and P.V. Oudemans, Rutgers University, Chatsworth, NJ (48)

ABSTRACT

In New Jersey, the time-honored tradition of clean culture blueberry production employs intense tillage and herbicide application to maintain a weed-free environment. Declining soil health is likely a direct result of this cultural practice in a long-lived perennial crop and has led to difficulty with the establishment of new blueberry plantings. The purpose of this study was to test several cover crop species in the low pH, sandy soils of a typical blueberry farm on the Outer Coastal Plain of New Jersey. In March of 2016, a randomized complete block design was set-up at a blueberry farm in Hammonton, NJ with 11 cover crop and mulch/compost treatments with six replications. Individual plots were 8.5 meters by 11 meters to accommodate approximately 35 blueberry bushes for an eventual study of blueberry establishment. The field utilized for this study had been dedicated to blueberry nursery production from 1975 to 2015. Baseline soil samples were collected in April of 2016 and succeeding samples have been collected every six months to monitor changes in soil organic matter content. Each cover crop is evaluated after planting to determine plant density. Additionally, total biomass production is measured at termination of the cover crop. Throughout the summer of 2017, soil volumetric water content was measured in each plot. Lastly, the influence of each treatment on weed species density and abundance was assessed in July and August 2016. Development of a greenhouse bioassay to monitor changes in soil health is underway. Soils collected from each plot in the study are potted and planted with cranberry cuttings (stolons). The bioassay takes place over an 8-week period and root weight of cranberry plants are measured. The use of cranberry cuttings, rather than blueberry, is advantageous because they are more uniform, root quicker, and are readily available all year. Another greenhouse bioassay is being developed to evaluate the potential impact of soil organic matter on soil suppressiveness to *Phytophthora cinnamomi*. In this study, rooted blueberry cuttings are grown in field soils from the cover crop trial and inoculated with *P. cinnamomi*. After a 12-week period, root weight is measured and percent inhibition of root growth is determined.

EVALUATION OF BICYCLOPYRONE IN WATERMELON. M.A. Cutulle* and H.T. Campbell, Clemson University, Charleston, SC (51)

ABSTRACT

Bicyclopyrone (BIR) is currently being evaluated in multiple horticultural crops including watermelon (*Citrullus lanatus*). BIR is a new HPPD (4-hydroxyphenyl-pyruvate dioxygenase) inhibitor that has improved grass activity relative to other HPPD-inhibiting herbicides. A field trial was conducted at the Clemson Coastal Research and Education Center (CREC) to determine the safety and effectiveness of BIR programs in watermelon applied pre-emergently (PRE) and the ability of chelated Fe to safen BIR applied post-emergently. The treatments consisted of BIR (50 g ai/ha), halosulfuron (26 g ai/ha), and fomesafen (210 g ai/ha) herbicide applied alone and in multiple combinations underneath and over the plastic. Fascination watermelon transplants were planted 7 days after the PRE treatments were applied. Six weeks after transplanting BIR was applied POST with and without Sequestrene, a chelated Fe product. Percent goosegrass (*Eleusine indica*) ratings taken 3 weeks after PRE application ranged from 65% control from halosulfuron applied over the plastic to 95% control with the halosulfuron, BIR, fomesafen combination under the plastic. Minimal watermelon injury was observed from any of the plots subjected to PRE treatments. Applications of BIR applied POST alone resulted in 15% injury while combining BIR with Sequestrene resulted in less than 10% injury. BIR has the potential to improve weed control when applied in multiple scenarios during watermelon production.

EVALUATION OF BICYCLOPYRONE IN SWEET POTATO. H.T. Campbell*, M.A. Cutulle, P. Wadl, and B. Ward, Clemson University, Charleston, SC (52)

ABSTRACT

Experiments were conducted in 2017 to determine the safety of bicyclopyrone (BIR) in sweet potato. Initial greenhouse experiments at the Clemson Coastal Research and Education Center (CREC) in Charleston, SC evaluated the response of Beauregard, Bunch, Sumor, Covington, LA07, Excel, and 04136 (an experimental bunched type variety) sweet potato varieties to BIR. BIR was applied pre-transplant to a sandy-loam soil cut with 25% river sand in round pots at 0, 50, 100, and 200 g ai/ha. Generally, the 50 g ai/ha rate of BIR did not have any phytotoxic effects on the varieties; however, weights of the plants trended to decrease when comparing the 50 g ai/ha rate to the untreated check. When accessing injury and bleaching at higher rates Beauregard trended to be the most tolerant while Sumor was the least tolerant to BIR. 04136 did not show any phytotoxic symptoms at the 1x rate. However, the greatest reduction in fresh weight relative to the untreated check when 200 g ai/ha of bicyclopyrone was applied was observed in the 04136 variety. A field trial was conducted at the Edisto Research and Education Center (EREC) to determine the value of BIR herbicide programs in Sweet Potato and determine if sequestrene a chelated Fe product increased the safety of BIR in sweet potato. BIR was applied at 50 g ai/ha alone and in combination with Dual Magnum at 800 g ai/ha, and Valor at 210 g ai/ha. All Dual Magnum treatments were applied two weeks post-transplant (POST) and all Valor treatments were applied PRE. A commercial PRE program utilizing PRE applications of two treatments of BIR at 50 g ai/ha also included a combination with Sequestrene at 224 g ai/ha, with 1 treatment applied PRE and 1 treatment applied POST. On. All weed control data and sweet potato injury data were collected and checked against a weed-free hand weeded treatment, as well as an untreated check treatment. Valor provided the greatest control of Palmer Pigweed, which was the most problematic weed at this location; however, combining bicyclopyrone with Valor resulted in a 10% increase in Palmer Pigweed control. The only injury observed in the trial was in plots treated with BIR, though the injury never exceeded 10%. Sequestrene did not significantly reduced BIR injury in Sweet potato. Based on this study, if registered BIR has the potential to be a supplementary weed management tool in sweet potato.

PRE-HARVEST CONDITIONING IN POTATO TO ENHANCE SKINNING RESISTANCE IN VIRGINIA'S EASTERN SHORE. R.A. Arancibia*, C.W. Cahoon, J. Jenrette, and H. Ferebee, Virginia Tech, Painter, VA (55)

ABSTRACT

Skin set is critical in redskin potato cultivars due to the unappealing off color scars and the new skin. We investigated pre-harvest conditioning with growth regulators and the time of chemical vine-kill before harvest to set the skin (adhesion) and maintain potato quality. In addition, we evaluated the potential internal disorders as a result of the treatments. Studies were conducted at the Eastern Shore AREC in a Bojac sandy loam soil with 'Dark Red Norland', 'Superior', and 'Red Norland'. The progression of skin adhesion was determined over time after treatment by the force required to snap the skin off (resistance to skinning) using a Halderson periderm shear tester. Resistance to skinning in response to pre-harvest treatment varied depending on chemical used and harvest time after treatment. Diquat (Reglone) is the most used treatment in commercial potato production and set the skin 2 to 3 weeks after treatments. Application of Ethephon, Actigard and Sharpen are promising alternatives that need further investigation. Minimal internal disorders were detected in these studies.

CROP TOLERANCE AND WEED CONTROL WITH PREEMERGENCE HERBICIDE COMBINATIONS IN HIGHBUSH BLUEBERRY (*VACCINIUM CORYMBOSUM*). F. dAmico*, B. Carr, and T.E. Besançon, Rutgers University, Chatsworth, NJ (57)

ABSTRACT

New Jersey produced 20 million kg of blueberries in 2016 at a farm value of $59 million (USDA 2017). Field studies were conducted in 2016 at a three commercial blueberry farms near Hammonton, NJ, to determine weed control and crop tolerance to various combinations of preemergence herbicides, including a recently registered premix of sulfentrazone and carfentrazone-ethyl. Herbicides were applied pre-budbreak and consisted of diuron at 1.4 kg ai ha^{-1} alone or mixed with either flumioxazin at 290 g ai ha^{-1}, mesotrione at 160 g ai ha^{-1}, or sulfentrazone at 210 g ai ha^{-1}. Additional treatments consisted in mesotrione at 160 g ai ha^{-1}, sulfentrazone at 210 g aiha^{-1} or sulfentrazone at 420 g ai ha^{-1}. All treatment also included oryzalin at 4.5 kg ha^{-1}. Herbicide programs containing flumioxazin or mesotrione provided greater horseweed control (100%) 5 weeks after treatment (WAT) than herbicide program containing sulfentrazone at 210 g ha^{-1} alone or tank mixed with diuron (85% and 75%, respectively). Increasing the rate of sulfentrazone to 420 g ai ha^{-1}resulted in greater horseweed control 5 WAT (94%). Mesotrione and flumioxazin alone or mixed with diuron provided greater toadflax control (>93%) 5 WAT than treatments that contained sulfentrazone at 210 g ai ha^{-1} alone or mixed with diuron (84%). Slender goldenrod control by all herbicide programs was between 73% and 88% 5 WAT. Programs that included mesotrione and flumioxazin maintained excellent horseweed control (>95%) 14 WAT, whereas programs containing sulfentrazone at 210 g ai ha^{-1} had 65% to 73% horseweed control. Sulfentrazone alone at 210 g ai ha^{-1} had significant lower toadflax control (71%) 14 WAT than when mixed with diuron (89%) or programs that included mesotrione or flumioxazin (>86%). Greater goldenrod control (>70%) 14 WAT was achieved with programs mixing mesotrione or flumioxazin with diuron than programs without diuron or relying on sulfentrazone at 210 g ai ha^{-1} (26% to 48%). However, increasing sulfentrazone rate to 420 g ai ha^{-1} helped to maintain higher goldenrod control 14 WAT (61%).

JUST WHAT DOES BICYCLOPYRONE BRING TO THE PARTY? L.L. Smith*, R.D. Lins, G.D. Vail, and T.H. Beckett, Syngenta, King Ferry, NY (58)

ABSTRACT

The HPPD-inhibiting herbicide bicyclopyrone has been developed for the corn weed control market as a component in active ingredient mixture products (Acuron, Acuron Felxi) and commercially launched in 2015. Mixtures with bicyclopyrone have shown improved weed control compared to products with similar active ingredients (Lumax, Lexar, Zemax). However, as a mixture component, little information is widely available regarding the activity of bicyclopyrone applied alone. This paper highlights the weed control benefits that bicyclopyrone provides when applied alone and in mixtures.

ASSESSING ZERO PALMER AMARANTH SEED RETURN IN A GRAIN SORGHUM-SOYBEAN ROTATION VERSUS CONTINUOUS SOYBEAN. M. Flessner*, S. Haring, S. Beam, and K. Bamber, Virginia Tech, Blacksburg, VA (59)

ABSTRACT

A zero-tolerance policy is currently recommended for Palmer amaranth (*Amaranthus palmeri*) management, but eliminating returns to the weed seed bank via hand weeding are costly. Alternatively, crop rotation to grain sorghum may be more economical, due to available herbicide options. Research was conducted to evaluate Palmer amaranth control and economics thereof in a grain sorghum-soybean rotation versus continuous soybean.

The experiment had two main factors: crop rotation and Palmer amaranth management program. Crop rotation was grain sorghum-soybean versus continuous soybean. Management programs varied by crop. The program for grain sorghum was atrazine at 1.5 kg ha^{-1} + *S*-metolachlor at 1.5 kg ha^{-1} + mesotrione at 0.09 kg ha^{-1} PRE. Four management programs were assessed in soybeans: no weed control (nontreated), a "low input" program of flumioxazin at 0.14 kg ha^{-1} PRE followed by (fb) glyphosate at 1.1 kg ha^{-1} POST, a "high input" program of sulfentrazone at 0.15 kg ha^{-1} + *S*-metolachlor at 1.4 kg ha^{-1} + metribuzin at 0.42 kg ha^{-1} PRE fb fomesafen at 0.27 kg ha^{-1} + glyphosate at 1.1 kg ha^{-1} + *S*-metolachlor at 1.4 kg ha^{-1} POST and a "zero-tolerance" program that was the "high input" program + hand-weeding after POST, prior to seed shed. The experiment was a split-plot randomized complete block with four replications. Plot locations were identical across years. Weed control, yield, and an economic analysis were assessed.

Due to a poor cropping year, no treatments were profitable in year one. However, results indicated that grain sorghum lost the least money, followed by the "low input," followed by the "high input" and nontreated, and "zero tolerance." Grain sorghum was the cheapest way to ensure zero Palmer amaranth seed return, due to the high cost of hand weeding.

Data analyses from year two indicate the grain sorghum-soybean rotation resulted in better Palmer amaranth control (77 versus 60%, respectively four weeks after POST; $p<0.001$), 15% greater yield ($p=0.030$), and was 26% more profitable ($p=0.030$) compared to continuous soybeans. Palmer amaranth counts were collected from nontreated plots in year two, and indicated that grain sorghum-soybean rotation reduced Palmer amaranth density by 3.1 fold ($p<0.001$) compared to continuous soybeans.

Combining the loss from year one and profit in year two, the grain sorghum-soybean rotation was 88% more profitable than continuous soybean ($p=0.007$) when assessed across all programs or 60% more profitable ($p=0.014$) when assessed across programs excluding the nontreated.

DIFFERENCES IN WEED COMMUNITY CHARACTERIZATION BETWEEN TWO SEED BANK ANALYSIS TECHNIQUES. T.A. Reinhardt* and R. Leon, North Carolina State University, Raleigh, NC (60)

ABSTRACT

Seed bank sampling and quantification techniques vary in their ability to describe weed diversity and density, so proper technique selection is critical for studying weed communities. The germinable seed bank technique is preferred over seed extraction because the latter is more time consuming. However, these two methods have only been compared using a few weed species and a relatively small number of samples. We used 204 weed seed bank samples to compare both techniques for weed density richness, evenness, and diversity. The two methods yielded dramatically different results. The germinable seed bank technique generally underestimated all parameters compared with the extraction technique. No clear trend on weed composition or frequency was observed regarding differences between weed seed bank techniques. This is the largest study of paired samples to test the comparability of these techniques to characterize weed communities, and the results imply that the strengths and pitfalls of each method need to match the intentions of the study. Thus, the germinable seed bank is a practical approach to compare treatments or conditions. However, the extractable seed bank is more useful to accurately quantify weed richness, diversity and density.

HALAUXIFEN-METHYL: SPECTRUM OF BURNDOWN WEED CONTROL. C.W. Cahoon*, A.C. York, M. Askew, M. Flessner, and T. Hines, Virginia Tech, Painter, VA (61)

ABSTRACT

Halauxifen-methyl is a member of the new arylpicolinate family of auxin herbicides developed by Dow AgroSciences. A pre-mix of halauxifen-methyl plus florasulam is currently labeled for use in wheat, barley, and triticale. Halauxifen-methyl alone is being marketed for horseweed control preplant burndown prior to planting corn, cotton, soybean, and other crops. Previous research has shown halauxifen-methyl effectively controls horseweed and henbit. However, little is known about its efficacy against many other common winter weeds. The objective of this study was to evaluate control of small and large horseweed and other common weeds encountered preplant burndown. Experiments were conducted near Jackson, NC, Ramseur, NC, three separate fields near Painter, VA (PL1A,PL1B, and PJ1), and two separate fields near Rocky Mount, NC (RM1 and RM2). Treatments were arranged in a randomized complete block design with treatments replicated 3 or 4 times. Treatments included halauxifen-methyl (0.004 lb ai A^{-1}), dicamba (0.25 lb ae A^{-1}), 2,4-D low rate (LR) (0.475 lb ae A^{-1}), 2,4-D high rate (HR) (0.95 lb ae A^{-1}), glyphosate (1.125 lb ae A^{-1}), halauxifen-methyl + glyphosate, dicamba + glyphosate, 2,4-D LR + glyphosate, and 2,4-D HR + glyphosate. Methylated seed oil at 1% V/V was included with halauxifen-methyl and halauxifen-methyl + glyphosate whereas nonionic surfactant was included with dicamba and 2,4-D when applied alone. Horseweed and cutleaf eveningprimrose were observed at 4 of 6 locations. Halauxifen-methyl controlled horseweed 74%. Dicamba (73%) and 2,4-D HR (73%) controlled horseweed similar to halauxifen-methyl whereas 2,4-D LR (53%) was less effective. Halauxifen-methyl, dicamba, 2,4-D LR, 2,4-D HR, and glyphosate controlled cutleaf eveningprimrose 8, 49, 84, 93, and 28%, respectively. Halauxifen-methyl effectively controlled henbit (90%), common vetch (85%) and yellow woodsorrel (98%). However, the herbicide was less effective against common chickweed (10%), mouse-ear chickweed (0%), cudweed (3%), curly dock (10%), field violet (0%), and wild garlic (0%).

WINTER GRAIN MANAGEMENT STRATEGIES IN A REDUCED-TILLAGE, ORGANIC SYSTEM: A WEED MANAGEMENT PERSPECTIVE. J.M. Wallace*, R. Champagne, and W. Curran, Cornell University, Geneva, NY (62)

ABSTRACT

[no abstract submitted]

COVER CROP IMPACT ON NO-TILL CORN YIELD IN A MULTI-SITE TRIAL. E.D. Sweep*, University of Maryland, College Park, MD (63)

ABSTRACT

Cover crops can affect the yield of the subsequent cash crop by altering N cycling and moisture dynamics. Corn (*Zea mays* L.) yield was measured at 16 sites across Maryland and Pennsylvania in an on-farm trial. This experiment is to be repeated for an additional 4 years; we present preliminary data from the first year (2017) here. Each site was a production grain farm with cover crops (primarily small grains, and especially cereal rye [*Secale cereale* L.]) in rotation following soybean (*Glycine max* L. Merr.); we compared yields in 2 subplots each within paired strips of cover crop and winter-fallow. Cover crops at spring termination represented a wide range of biomass, 3.09+/-3.25 Mg ha^{-1}; termination dates ranged from 22 March to 25 May. Grain yields were analyzed using a linear mixed model with fixed effects of treatment (winter-fallow or cover crop) and random effects for farm and subplot. We found that yields following cover crop decreased 0.49 Mg ha^{-1} (+/-0.16, $p<0.01$). At some farms, cover crops increased corn yield; the effect of cover crops within site ranged from -1.22 to 0.713 Mg ha^{-1}.

FEHEDTA EFFICACY ON SEEDLING AND MATURE WEEDS. J. Neal* and C. Harlow, North Carolina State University, Raleigh, NC (65)

ABSTRACT

FeHEDTA is currently labeled as a biopesticide for broadleaf weed control in turfgrass. Data has shown that multiple applications are necessary to control most broadleaf weeds. When applied twice or three times at fourteen day intervals, essentially complete control of seedling *Lamium amplexicaule* and *Stellaria media* seedlings has been reported. Additionally, when applied to seedling broadleaf weeds grown in containers excellent control of *Oxalis stricta* has been reported, but variable efficacy on several other broadleaf weeds was observed. In this experiment, three doses of FeHEDTA were applied to seedling and mature dicot weeds common to container nursery crop production. One-liter pots filled with a pine bark substrate were surface seeded with *Oxalis dillenii, Lamium amplexicaule, Stellaria media, Senecio vulgaris, Cardamine flexuosa,* and *Eclipa prostrata*, each in separate pots. Pots were overhead irrigated outdoors. FeHEDTA (as Fiesta herbicide) applications to seedling and mature plants were compared. FeHEDTA was applied at 25, 50 or 100 oz product per 1000 ft^2 in 50 GPA of carrier volume. Seedling and mature *Senecio* and *Eclipta* seedlings were controlled >95% by 50 oz/1000^2FeHEDTA. For all other weeds there were differences in efficacy between plant ages. Seedling oxalis was well controlled with $\geq$ 50 oz/1000^2 but mature oxalis plants exhibited leaf necrosis but recovered rapidly resulting in <25% control. Conversely, older *Lamium* and *Cardamine* plants were better controlled than young plants. These surprising results illustrate that additional research on the efficacy of FeHEDTA is needed.

COMMON LESPEDEZA CONTROL IN COOL-SEASON TURFGRASS. J. Craft* and S. Askew,
Virginia Tech, Blacksburg, VA (66)

ABSTRACT

Common lespedeza (*Lespedeza striata*) is an aggressive mat-forming summer annual weed in turfgrass systems. Common lespedeza is commonly found in areas of thin, compacted and poorly fertilized turfgrass. As common lespedeza matures in late summer, the stems become woody which enhances its tolerance to drought, heat and low mowing heights. Multiple applications of premixed products that contain 2,4-D, MPCA, mecoprop, and dicamba are effective if applied early in the season. However, as summer progresses these products lose their effectiveness due to lespedeza becoming more mature and developing woody stems. Research has shown products such as fluroxypyr and triclopyr control common lespedeza with a single application, but during the summer these products could be slightly injurious to cool-season grasses. Therefore, the objective of this research was to evaluate common lespedeza response and cool-season turfgrass response to several commercially-available broadleaf herbicides and some experimental products. Several trials were conducted in Blacksburg, VA to evaluate different herbicides and combinations for lespedeza control. Trials were also conducted to study tall fescue and fine fescue tolerance. Speedzone (carfentrazone, 2, 4-D, mecoprop, dicamba), Speedzone Southern (carfentrazone, 2,4-D, mecoprop, dicamba), and MSMA had the quickest rate of activity compared to other products. However, Speedzone Southern caused discoloration and stunting initially, but regrowth began 14 days after treatment (DAT). Speedzone had the quickest rate of activity compared to other products; however, by the conclusion of the trial, control had dropped to 65%. While MSMA controlled common lespedeza greater than 90%, it injured tall fescue 75%. 4-Speed XT (2, 4 D, triclopyr, dicamba, pyraflufen-ethyl) controlled common lespedeza greater than 98% while causing little to no injury to tall fescue. EH1587 (halauxifen-methyl, fluroxapyr, dicamba) controlled common lespedeza greater than 90% while causing little to no injury to tall fescue or fine fescue. Data suggest EH1587 can selectively control common lespedeza in tall fescue or fine fescue. It was observed that products containing triclopyr or fluroxypyr controlled common lespedeza greater than common four-way broadleaf herbicides.

NUTSEDGE CONTROL IN NURSERY FIELDS. H.M. Mathers*, Mathers Environmental Science Services, LLC, Gahanna, OH (68)

ABSTRACT

Yellow nutsedge, *Cyperus esculentus* L., is a member of the Cyperaceae (sedge family) of monocotyledonous flowering plants reproducing primarily from tubers. Tubers break dormancy in winter and germinate in the spring. Past studies with Sedgehammer (Halosulfuron-methyl) (Group 2) (Gowan Company, Yuma, AZ) (0.05 lb a.i/ac) and FreeHand® 1.75G (Dimethamid-p 0.75% (Group 15) + Pendimethalin 1% (Group 3) (BASF Corp., Research Triangle Park, NC) (3.5 lb a.i/ac) applied on March 27, 2009 in Lake County, Ohio, before emergence, provided control through July or 3 months after treatment (MAT). Mature tubers are known to be unaffected by glyphosate (Wright and Vargas, 2003). The objective of this study was to evaluate Vestis™ (Siloxane Polyalkyleneoxide Copolymer and Polyalkyleneoxide) (Precision Laboratories, Waukegan, IL) an organosilicone wetting agent, after emergence, for possible improved penetration through the cuticle and corresponding increase and/or extension of efficacy with halosulfuron-methyl or glyphosate containing products. Additionally, evaluations of a new preemergence herbicide, V-10233, not yet registered in the ornamental market (Valent USA, Corporation, Walnut Creek, CA) [(Flumioxazin (33.5%) (Group 14) + Pyroxasulfone (42.5%) (Group 15)] with and without halosulfuron-methyl or glyphosate and/or Vestis™ were conducted. Nufarm ProSedge® (Halosulfuron-methyl) Nufarm Americas, Inc., Burr Ridge, IL and Rodeo ® (glyphosate isopropylamine salt 53.8% (Dow AgroSciences, Indianapolis, IN) were selected for this study. Lontrel® T&O (Dow AgroSciences) (Clopyralid 40.9%) was also evaluated and provided little control post emergence even with Vestis™. Trials were initiated on May 31, 2017 with evaluations at 3, 9 and 14 weeks after treatment (WAT) in Lenox, MI and on August 9, 2017 with evaluations at 5 WAT in Grand Haven and West Olive, MI. In the three trials, Nufarm ProSedge® + Vestis™ provided greater efficacy and extended efficacy in Lenox, MI to 14 WAT versus ProSedge® applied alone. Rodeo was only tested in combination with Vestis™ and V-10233 in the August 9th studies and provided 100% control (rating of 10) at 5WAT in both locations. Only in the Grand Haven, MI trial was Nufarm ProSedge® + Vestis™ + V-10233 tested. At this site it provided statistically similar efficacy to Rodeo + Vestis™ + V-10233 (9.8). Further evaluations will resume in early spring with the August 9th locations and preemergence applications in December, 2017 and March, 2018 are planned for two additional MI nurseries.

SPRING EVALUATIONS OF PHYTOTOXICITY AND EFFICACY OF DORMANT PREEMERGENCE APPLICATIONS TO NURSERY CONTAINERS. H.M. Mathers*, Mathers Environmental Science Services, LLC, Gahanna, OH (69)

ABSTRACT

Liquid preemergence herbicides can be applied more uniformly; thus, reducing waste and expense. Liquid applications on dormant container stock utilize staff in winter, a "down-time" labor-wise, and ensure applications are completed before conflicting spring operations take precedence. Unfortunately, there are no preemergence herbicides registered for use in enclosed overwintering polyhouses and consequently few studies have been conducted comparing the phytotoxicity of liquid vs granular formulations for this use. The objectives of this study were to evaluate the phytotoxicity of very early spring herbicide applications that would coincide with venting via cutting holes in the polyhouses, opening of the end-doors, and/or other means of stock exposure. To this end, four liquid formulations currently being used in the industry (off-label) including, SureGuard 6oz, Marengo SC 7 oz/ac, Marengo SC 15 oz/ac and Gallery/Dimension (1.0+0.5lb a.i.) were applied over-the-top (OTT) and compared to a common granular (off-label) OTT Snapshot 200 #/ac, and an untreated control. Additionally, evaluations for extended efficacy, into early July, were conducted. Ten species of common container grown plants were evaluated at two commercial nurseries one in Huron and another in Perry, OH with over the top applications on dormant growth (March 20, 2015 or April 9, 2015), respectively. Five of ten species showed no injury from any the dormant treatments including at Huron, OH, *Berberis thundbergii* 'Arto Rose Glow' (1 gal.) and *Rhododendron* 'Nova Zembla' (1 gal.), and at Perry, OH, *Juniperus scopulorum* 'Blue Arrow' (1 gal), *Taxus Xmedia* (4"), and, *Buxus* 'Green Gem' (1 qt.). However, *Spirea japonica* 'Magic Carpet,' (Huron, 1 gal), *Spirea bumalda* 'Goldflame' (Perry, 3 gal), *Viburnum* X Juddi (Huron, 1 gal.), *Ilex Xmeserve* 'China Boy' (Huron, 1 gal) and *Ilex Xmeserve* 'Blue Prince' (Perry, 1 gal.) did have injury and/or died when growth resumed in the spring. The OTT dormant herbicide applications were considered related to increase winter injury and/or death, as no controls died or appeared injured. Two subsequent herbicides were applied, per normal practices, in summer and fall. Evaluations were conducted at 6 weeks after the treatment (WAT), 8, 10, 12, 18, 22, 25, and 28 (WAT). At the conclusion of the trial, eight months after initiation, final heights and plant diameters (an average of two perpendicular measurements W1 and W2, and destructive samples were conducted. From these measures growth index (GI), Fullness (F) and root and shoot dry weight, (RDW) and (SDW), respectively, were completed. A positive correlation with injury from the dormant applications could be made to the final growth index (GI) values. Growth index was calculated as GI=Pi (Ht)(r^2), where Ht. was final height, r was half of the average of W1+W2 and Pi was "p". Of the three genera *Spirea*, *Ilex* and *Viburnum*, Marengo SC applied at 15 oz/ac caused the greatest winter injury for *Spirea* and *Ilex*. Marengo SC 7.5 oz/ac caused similar injury to 15 oz/ac with Spirea. The Gallery + Dimension caused the greatest injury on *Viburnum*. The Marengo SC label does read whether applied dormant or to actively growing plants, "spray must be directed at the base of the plant or away from the plant." Snapshot caused the least injury for all species.

CARRYOVER OF RESIDUAL HERBICIDES TO VARIOUS COVER CROP SPECIES. M. Flessner*, K. Pittman, S. Beam, L. Rector, K. Bamber, R. Randhawa, and S. Haring, Virginia Tech, Blacksburg, VA (70)

ABSTRACT

Increasing herbicide resistance has placed renewed focus on the use of residual herbicides. These herbicides have the potential to remain active in the soil and result in carryover injury to cover crops planted after harvest. Limited research and information are available to inform farmers as to which cover crop species are sensitive to which residual herbicides. Therefore research was conducted to evaluate 30 herbicides for carryover to 10 cover crop species. The research was conducted as a split-plot randomized complete block design with 4 replications. The main plot was herbicide, which was applied in early-July if the herbicide could be applied in-crop or in early-June if must be applied prior to crop planting. Sub-plots were cover crop species, which were drilled in early-September. Land was fallow from June until cover crop planting, and weeds were controlled with glyphosate as necessary. The experiment was conducted at two sites in Virginia: Suffolk in a loamy sand and Blacksburg in a loam. The experiment was conducted in 2016 and 2017 for a total of 4 site-years. Data collected included visible cover crop injury relative to a nontreated check within cover crop species on a 0 (no injury) to 100% (complete necrosis) scale 3 and 6 weeks after planting (WAP). Above ground biomass was collected from 1m of row for all species 6 WAP. Data were analyzed using JMP Pro 12 by determining significant effects (alpha=0.1) with ANOVA and subsequent means separation using Fisher's Protected LSD. Visible injury was influenced by year and location, indicating that soil-type and environmental factors such as temperature and rainfall influence herbicide persistence. Across herbicides and species Suffolk in 2016 resulted in the greatest injury of 6.5% 3 WAP and 6.3% 6 WAP. Other site-years resulted in <2% injury. Mostly, no visible injury was detected, but individual herbicides did result in injury to various cover crop species in certain cases. Injury, when observed, was typically 10 to 20%, but a maximum of 40% injury was observed. These data largely corroborate the 2017 Mid-Atlantic Field Crop Weed Management Guide.

GROWTH EFFECTS OF SORGOLEONE (SORGHUM ROOT EXUDATE) ON DIFFERENT WHEAT AND WEED SPECIES. M. Bansal* and W.J. Everman, North Carolina State University, Raleigh, NC (71)

ABSTRACT

Sorghum in known to produce allelochemical called 'Sorgoleone'. Plants can produce these chemicals either by roots when they are still alive or by dead decaying matter and are known to have negative impact on weeds and following crops. There are concerns about sorghum affecting following winter wheat growth when grown in rotation in North Carolina. Lab studies were conducted in 2015 to evaluate the impact of sorgoleone on growth of wheat and different weed species. Seeds of wheat (Shirley) and four weed species, large crabgrass, Italian ryegrass, velvetleaf, and sicklepod were pre-germinated and then transferred to 20x100mm petri dishes treated with varying concentrations of sorgoleone. Sorgoleone was applied @ 0 (control), 25, 50, 100, 150, 200, and 300 μg ml^{-1}. 10 days after placing seeds on the petri dishes, growth was measured in terms of shoot length. Significant sorgoleone treatment effects were observed for shoot growth when pooled over species. Shoot length was reduced at higher rates of sorgoleone compared to control. Wheat shoot length was not significantly affected by sorgoleone concentration. Velvetleaf shoot length was lower at all concentration compared to control. At higher rates of sorgoleone, large crabgrass, Italian ryegrass, and sicklepod growth was reduced when compared to lower rates. Preliminary analysis suggests that sorgoleone has a negative impact on growth of weed species, however wheat is not impacted.

WHAT'S NEW. S.A. Mathew*, D.L. Bowers, A.S. Franssen, B. Minton, and M. Schraer, Syngenta, Gaithersburg, MD (72)

ABSTRACT

[no abstract submitted]

IMPACT OF PREEMERGENCE HERBICIDE APPLICATIONS ON CAROLINA REDROOT (*LACHNANTHES CAROLIANA*) CONTROL AND CRANBERRY (*VACCINIUM MACROCARPON*) FRUIT QUALITY. T.E. Besançon* and B. Carr, Rutgers University, CHATSWORTH, NJ (73)

ABSTRACT

New Jersey produced 29.6 million kg of cranberries in 2016 at a farm value of $28 million (USDA 2017). New Jersey production is concentrated in the Pine Barrens coastal plain where sandy acidic soils are optimal for cranberry. The perennial nature of cranberry predisposes the crop to a diversity of weed species ranging from herbaceous weeds to woody perennial species. Carolina redroot [*Lachnanthes caroliniana* (Lam.) Dandy] is a perennial herbaceous weed species member of the *Haemodoraceae* family. Its proliferation and the lack of efficient control strategies has been an increasing source of concern for New Jersey cranberry growers. Information regarding herbicidal control of Carolina redroot is extremely limited and restricted to blueberry production (Myers et al. 2013). A study was initiated in 2017 in order to evaluate Carolina redroot control and crop response following a spring application of preemergence herbicide at various labelled rates. Evital 5G (norflurazon) was applied at 560, 1,120, 2,240, and 4,480 g ai ha^{-1}, Devrinol DF-XT (napropamide) at 6,720 g ai ha^{-1}, and Casoron 4G at 2,240 and 4,480 g ai ha^{-1}. Carolina redroot control 8 WAT was 63% with Devrinol DF-XT, 85% and 99% with Casoron at 2,240 and 4,480 g ai ha^{-1}, respectively. Control remained over 70% 15 WAT with Casoron, regardless of rates, but decreased to 39% with Devrinol. Evital never provided more than 5% Carolina redroot control, regardless of rate applied. Carolina redroot density was 480 plants m^{-2} 15 WAT in the untreated check but only 130 plants m^{-2} with Casoron at 4,480 g ai ha^{-1}, and averaged 345 plant m^{-2} with Casoron at 2,240 g ai ha^{-1} and Devrinol. Carolina redroot dry biomass 15 WAT was not significantly reduced with Evital at rates ranging from 560 to 2,240 g ai ha^{-1}, but decreased by 27% at 4,480 g ai ha^{-1}. Devrinol as well as Casoron at 2,240 and 4,480 g ai ha^{-1} significantly reduced dry biomass 15 WAT by 63, 68 and 90%, respectively. Cranberry injury, in the form of leaf chlorosis, was noted with the all herbicides, peaking 12 WAT with 11% for Casoron at 4,480 g ai ha^{-1}, 6% for Casoron at 2,240 g ai ha^{-1} and Devrinol, and not exceeding 3% with Evital. Casoron, regardless of rate, caused significant stunting with 13 to 15% 12 WAT whereas Evital and Devrinol did not impact cranberry growth. If chlorosis disappeared 15 WAT, stunting induced by Casoron persisted longer with 4 and 9% for at 2,240 and 4,480 g ai ha^{-1}, respectively.

INTEGRATION OF CEREAL RYE AND HERBICIDES FOR WEED MANAGEMENT IN WATERMELON. K.M. Vollmer* and M. VanGessel, University of Delaware, Georgetown, DE (75)

ABSTRACT

Weeds cause significant yield loss in watermelon production systems. Commercially acceptable weed control is difficult to achieve even with heavy reliance on herbicides. These herbicides often do not provide season-long weed control between plastic rows. In addition, several weed species in this region have become resistant to commonly used herbicides. Cover crops have been adopted in many production systems to improve weed control. The objective of this study was to evaluate spring-seeded cereal rye cover crop management with pre- or post-transplant residual herbicides for weed management between plastic mulch. The study was a two-factor factorial with main effects of cover crop termination timing and additional weed control (residual herbicides, weed-free check, and no additional treatment). Beds were formed and cereal rye (134 kg ha^{-1}) seeded one month prior to watermelon transplanting. Cover crops were terminated 3 and 5 weeks after transplanting (WATrplt) with clethodim (136 g ha^{-1}) plus a nonionic surfactant (0.25% v/v). Residual herbicide treatments were applied at transplanting and 2 WATrplt. Residual herbicides were halosulfuron (15 g ha^{-1}) plus *s*-metolachlor (1,346 g ha^{-1}). Cover crop biomass was collected prior to each termination date by removing four 0.25 m^2 quadrats from each plot then dried to a constant weight. Weed density was taken 0, 2, and 4 WATrplt using at least three 0.25 m^2 quadrats in each plot, and weed biomass collected 4 WATrplt using the same quadrats then dried to a constant weight.

There was no difference in cereal rye biomass collected prior to each termination timing. At transplanting, Palmer amaranth density was lower in cereal rye plots, but there were no differences in common lambsquarter density. Common lambsquarter and Palmer amaranth densities were lower in cereal rye plots at 2 WATrplt, regardless of herbicide application. At 4 WATrplt Palmer amaranth density was lower in cereal rye plots terminated late, compared to plots without cereal rye. Common lambsquarter density was lower in cereal rye plots, regardless of termination timing or herbicide application. Palmer amaranth biomass was at least 32 g m^{-2} in cereal rye plots and at least 158 g m^{-2} in plots without a cereal rye cover. Common lambsquarter biomass was lower in cereal rye plots (2 g m^{-2}) compared to plots without cereal rye (13 g m^{-2}). These results show that between row cover cropping can help to reduce weed density. However, the level of control was not satisfactory as plant density and biomass increased throughout the growing season, suggesting additional control measures are warranted. Additional studies are needed on the best methods to integrate cover cropping and herbicides for weed control in watermelon.

MOSS AS AN INCREASING WEED PROBLEM IN CRANBERRY BEDS. K.M. Ghantous* and H.A. Sandler, UMass Cranberry Station, East Wareham, MA (77)

ABSTRACT

Moss is an emerging weed issue in cranberry growing regions including Massachusetts, Wisconsin, the Pacific Northwest, and Quebec. Although present in cranberry for decades as a minor weed, it has not been considered a management priority in the past. Growers have been adopting practices such as improved drainage and water management that should discourage moss growth over the last decade, however growers have recently begun to perceive moss as more widespread and difficult to control. Of growers surveyed at the 2016 UMass Cranberry Management Update Meeting, 67% reported having moss on their farms (n=100), 41 % said they felt it was more common than it was 5 years ago (n=85), and 41% said they considered moss to be a problematic weed (n=96). Although the most commonly recognized and prevalent cranberry weed mosses are haircap (*Polytrichum commune*) and sphagnum (*Sphagnum* spp.), a recent sampling made in June of 2017 of State Bog in East Wareham found at least three additional moss species present (*Aulacomnium palustre, Ceratodon purpureus,* and *Entodon seductrix*).

Traditional control measures, such as spot treatment of moss patches with high rates of iron sulfate, are not adequate for managing large infestations. Recent work screening registered herbicides on newer cranberry cultivars has confirmed that currently registered herbicides, such as dichlobenil (Casoron 4G, 4% a.i.) at 67.3 kg ha^{-1}, norflurazon (Evital 5G, 5% a.i.) at 89.7 kg ha-1 applied in either the spring or the fall, napropamide (Devrinol 2-XT, 22% a.i.) at 42.1 L ha^{-1}, quinclorac (QuinStar 4L, 40% a.i.) at 2 applications 0.61 L ha^{-1}, or mesotrione (Callisto, 40% a.i.) at 2 applications 0.58 L ha^{-1} provided little to no control of moss.

Over the past four years, we screened several products including acetic acid (20% horticultural strength), an herbicidal soap (22% ammoniated soap of fatty acids), Moss-aside (22% potassium salts of fatty acids), Impede (an insecticide with 49% potassium salts of fatty acids), Oxidate (5.34% hydrogen peroxide), iron sulfate (a feed-grade powdered form of 92% ferrous sulfate with 30% elemental iron) applied via drop spreader, iron sulfate (a feed-grade powdered form 92% ferrous sulfate with 30% elemental iron) dissolved in water and applied by chemigation, Moss-out Liquid (a 35% ferric sulfate with 9.75% elemental iron product), Scotts Moss Control Granules For Lawns (17.5% ferrous sulfate), and FeRROMEC (liquid turf product 15% Urea Nitrogen, 3% Combined Sulfur, 6% Iron). Of these products, the only ones that injured moss and were not injurious to cranberry vines were the powdered iron sulfate (both applied as a powder and also dissolved in water), Scotts Moss Control Granules (also iron sulfate), and Moss-aside (22% potassium salts of fatty acids).

Several herbicides not currently registered for use on cranberry were screened between 2015 and 2017 for possible efficacy against moss: carfentrazone (Aim); applied before cranberry budbreak), flumioxazin (Chateau), aminocyclopyrachlor (Method), fomesafen (Reflex), halosulfuron (Sandea), sulfentrazone (Zeus), and pyroxasulfone (Zidua). Flumioxazin and sulfentrazone both demonstrated good crop safety and moss control.

PROFESSIONAL DEVELOPMENT FOR UNIFYING RESISTANCE MANAGEMENT EDUCATION FOR NORTHEAST SPECIALTY CROPS. H.A. Sandler*, K.M. Ghantous, and L.G. McDermott, UMass Cranberry Station, East Wareham, MA (78)

ABSTRACT

Selection and use of available pesticides, with varied modes of actions (MoA), must be correct and judicious to forestall or avoid resistance development. Growers currently receive management information from various sources (e.g., Extension, pest advisers), but the information delivered is not necessarily consistent and comprehensive. Northeastern Extension and agricultural industry personnel need education to properly provide pest management and resistance management (RM) advice to specialty crop growers. This initiative, funded by a NE-SARE Professional Development Program grant, is creating and coordinating a unified approach to deliver RM education to Northeastern producers. Four "train-the-trainer" webinars, augmented by a Moodle resource platform, have been produced and uploaded to YouTube. We produced a core module (slide presentation to be utilized in outreach workshops) that is available at: http://scholarworks.umass.edu/cranberry_outreach_resistance/1/. We have also produced an educational video for reinforcement of RM principles. We will distribute surveys to capture changes in growers' knowledge and behavior after attending workshops where educators present RM information and will report on the results in the final report submitted to SARE.

The main objective for the project is that, after participating in the multiple facets of this grant, Extension and agricultural industry personnel from Northeastern states develop crop-specific training modules and materials to transfer this knowledge to Northeast specialty crop growers. The full survey results from assessing the resistance management education and experience of educators and growers in the Northeast can be found at: http://scholarworks.umass.edu/cranberry_research_repts/14.

We have been in contact with a small subset of individuals throughout the Northeast region to stay informed about workshops in their area and commodity. We will use this short list of interested individuals to help us keep track of RM training sessions over the next year. We will also engage this Short List to promote the use of the survey by educators. Twenty-eight (28) people, who registered for the webinar series, indicated that they are interested in obtaining a Certificate of Completion (COC), which requires attending at least 2 of the 4 webinars, conducting one-on-one or workshop education and participating in the verification portion of the grant. We anticipate that the tools development via this professional development grant will enhance and supplement education tools currently available (e.g. WSSA herbicide resistance modules) in the disciplines of entomology, plant pathology, and weed science.

COMPARISONS OF PRE-EMERGENCE APPLICATIONS OF SULFENTRAZONE, CARFENTRAZONE AND GLUFOSINATE FOR WEED CONTROL IN WILD BLUEBERRY (*VACCINIUM ANGUSTIFOLIUM*) FIELDS. D.E. Yarborough* and J. D'Appollonio, University of Maine, Orono, ME (79)

ABSTRACT

Sulfentrazone and carfentrazone (Zeus Prime XC, Group 14) and glufosinate (Rely 280, Group 10) have recently received a blueberry label, but we have little information on their efficacy for weed control in wild blueberries (*Vaccinium angustifolium* Ait.) in Maine. In spring 2017, trials were conducted on commercial wild blueberry fields in the towns of Union, Jonesboro and Wesley to encompass a range of weed and soil conditions. Treatments included an untreated check, Zeus alone, and Zeus with Rely in split treatments with hexazinone on one half; and Rely, hexazinone, terbacil, rimsulfuron and diuron in combinations, which were not split. All sites were evaluated in June and July for wild blueberry cover and phytotoxicity, and broadleaf and grass cover. Zeus application resulted in high initial injury to wild blueberry and therefore had the lowest initial cover among treatments; the addition of Rely and/or hexazinone did not increase the damage compared to Zeus alone. By the time of the second evaluation in July this injury was not as apparent, as the blueberry plants had recovered. Zeus phytotoxicity at Wesley was half that of the other sites. Wesley had the heaviest soil and highest organic matter, so it appears that on this site Zeus was tied up and so was less active. Although Wesley had less broadleaf weed control with Zeus alone compared to the other sites, in combination with hexazinone it was similar to the other sites and far better than the hexazinone combination treatments alone. Initial overall broadleaf weed cover in Union was low, and the treatments kept it low over time. Jonesboro had more broadleaf weeds initially, and the amount of weed suppression was increased with the addition of Zeus and Rely with hexazinone. In Wesley, although percent weed cover in the check was similar to Jonesboro in June and July, and the Zeus treatments responded similarly to the other sites over time, broadleaf weed cover in the Rely tank mix treatment in both June and July resulted in broadleaf weed cover that was higher than the untreated check. It may be that the Zeus rate used in this trial, 12.5 oz/a may be higher than necessary to control weeds on sandier sites. Phytotoxicity to wild blueberry was significantly lower in Wesley, indicating that some of the Zeus was being tied up in the heavier soil, but broadleaf weed control remained comparable to the other sites. This trial also confirms that soil composition can result in marked differences in weed responses when using soil applied herbicides. The heavier soil in Wesley prompted a release of broadleaf weeds in the Rely tank mix and hexazinone combination treatments instead of a reduction, so the addition of Zeus is needed on sites with heavier soils in combination with products such as terbacil, hexazinone and diuron as it resulted in much better control than the use of any of these combinations alone.

AUTHORITY MTZ: IS THERE A FIT IN PROCESSING TOMATOES? D. Lingenfelter*, M. VanGessel, and B. Scott, Penn State, University Park, PA (80)

ABSTRACT

Field studies were conducted in 2017 at two locations, Rock Springs, Pennsylvania and Georgetown, Delaware, to examine Authority® MTZ in transplanted tomato (*Solanum lycopersicum*, var. 'Heinz 3406') to determine its impact on crop injury and yield as well as effectiveness on annual weed control. PPI treatments were evaluated and included the Authority MTZ premix of sulfentrazone + metribuzin at five rates (0.169 lb ai/A, 0.225, 0.28, 0.338, and 0.45); and napropamide (1 lb) plus metribuzin (0.188 lb). A POST application of rimsulfuron (0.0313 lb) + metribuzin (0.14 lb) plus necessary adjuvants occurred in Delaware. Visual weed control evaluations were taken periodically throughout the growing period. Tomato yield data and crop phytotoxicity ratings were also collected. Small-plot studies were arranged in a randomized complete block design with three replications.

In Pennsylvania, the sulfentrazone + metribuzin treatments caused crop injury (25, 27, 32, 52, and 58%) respective to the rate from lowest to highest at 27 DAA. By 68 DAA, injury dropped to 12, 15, 18, 24, and 27%, respectively. The napropamide plus metribuzin treatment did not cause any crop injury. Yield was not taken at the Pennsylvania location. Similar trends were evident in Delaware, in which the sulfentrazone + metribuzin treatments caused crop injury (17, 19, 25, 50, and 53%) respective to the rate from lowest to highest at 27 DAA. At 50 DAA, injury dropped to 0, 5, 13, 14, and 21%, respectively. The napropamide plus metribuzin treatment caused no more than 3% crop injury. Yield data did not correlate well with injury. The untreated plots yielded 19.68 tons/A total fruit (red and green); while the sulfentrazone + metribuzin treatments produced 24.73, 23.90, 19.42, 19.16, and 22.42 tons/A, respectively. The napropamide plus metribuzin treatment yielded 23.96 tons.

In Pennsylvania, all sulfentrazone + metribuzin treatments at mid-season provided: 92 to 94% large crabgrass (*Digitaria sanguinalis*) control; 88 to 95% common lambsquarters (*Chenopodium album*) control; 84 to 96% redroot pigweed (*Amaranthus retroflexus*) and common purslane (*Portulaca oleracea*) control; and 93 to 98% eastern black nightshade (*Solanum phycanthum*) control. Napropamide plus metribuzin provided 60 to 89% control of the above species. In Delaware, the sulfentrazone + metribuzin treatments at mid-season provided: 83 to 87% large crabgrass (*Digitaria sanguinalis*) control; and 85 to 97% control of Palmer amaranth (*Amaranthus palmeri*). While the napropamide plus metribuzin treatment provided 77 and 91% control of large crabgrass and Palmer amaranth, respectively.

In summary, the Authority MTZ pre-packaged herbicide mix does have a fit in transplanted processing tomato production. However, the higher use rates might cause crop injury in certain soil types and under certain environmental conditions. The benefit of this herbicide product will be its ability to control certain difficult to control weeds such as eastern black nightshade, common lambsquarters, pigweed species, and likely annual morningglory that can be common in tomato crops. Also, the sulfentrazone portion of the product provides a unique herbicide mode of action (group 14) usually not used in tomato and can also help manage herbicide resistant weed biotypes. Additional testing of this herbicide would be useful to get a better understanding of its utility in this crop. Furthermore, crop injury might possibly be decreased if the herbicide was not incorporated.

EVALUATING HERBICIDE PROGRAMS IN TABLE BEET. D.E. Telenko* and J.R. Kikkert, Cornell University, East Aurora, NY (81)

ABSTRACT

Field trials in table beet (*Beta vulgaris* subsp. *vulgaris* L) were conducted in 2016 and 2017 to evaluate efficacy and crop tolerance to pre-emergence and post-emergence herbicide programs. Trials were established at the H.C. Thompson Research Farm in Freeville, NY on May 12, 2016 and at the NYSAES in Geneva on May 13, 2017. Varieties 'Red Ace' and 'Ruby Queen' were planted on 38.1 cm rows with 1.27 cm in-row spacing in Freeville and Geneva, respectively. In both years, an untreated check and 13 different herbicide treatments were evaluated. Treatments were arranged in a randomized complete block design with three replications. All plots were treated with metolachlor (Dual Magnum) at 0.72 kg ai ha^{-1} immediately after planting. There were six different concentrations of ethofumesate (Nortron) applied pre-emergence 2.13, 1.57, 1.05, 0.87, 0.70, and 0.53 kg ai ha^{-1}. The remaining treatments have both pre-emergent applications of metolachlor at 0.72 ai ha^{-1} and ethofumesate at 1.06 kg ai ha^{-1} followed by post-emergence herbicides applied alone or in combination and included the following: ethofumesate at 0.18 kg ai ha^{-1}, phenmedipham (Spin-Aid) at 0.27 1.06 kg ai ha^{-1} in 2016 and 0.09 kg ai ha^{-1}in 2017; triflusulfuron (UpBeet) at 0.017 kg ai ha^{-1}; clopyralid (Stinger) at 0.21 kg ai ha^{-1}; ethofumesate at 0.18 kg ai ha^{-1} + clopyralid 0.21 kg ai ha^{-1}, and a combination of all four products. The trial was designed to determine what each herbicide brings to a weed management program and help identify optimum programs for overall table beet weed management. Evaluation in both years included crop injury and weed control. Yield data was collected in 2016, but not in 2017 as the trial was terminated prior to crop maturity due to the extreme wet conditions that flooded the field.

No significant differences were detected between herbicide pre-emergence treatment and untreated control on % stunting, % chlorosis, and % stand reduction on 1 Jun on 'Red Ace' beet. Post-emergence treatments were applied on 21 Jun and on 8 Jul 14.3% dead plants were document in the phenmedipham treatment as compared to none in the untreated. Dead plants were also noted in the triflusulfuron treatment and pre-ethofumesate at 2.13 kg ai ha^{-1}, but these were not significantly different from untreated. No significant difference in yield was noted in 2016. No significant differences in % chlorosis, % stunting, or % stand reduction between herbicide programs and untreated control on 'Ruby Queen" beet were found in the 2017 trial.

In 2016, the prominent weed species included field bindweed (*Convolvulus arvensis* L), smartweed, lambsquarters (*Chenopodium album* L.), and redroot pigweed (*Amaranthus retroflexus* L.). On 1 Jun all programs had greater than 99% overall weed control compared to untreated, but by 20 Jun loss of weed control was noted in the metolachlor only program, and this continued through the rest of the season, as expected. It was also noted on 20 Jun that metolachlor + ethofumesate at 0.87 kg ai ha^{-1}was significantly less than metolachlor + ethofumesate at the full rate of 2.13 kg ai ha^{-1}. On 8 Jul all programs had greater than 94% weed control over the untreated, except metolachlor alone and metolachlor + ethofumesate 0.53 kg ai ha^{-1}. Weed species that were breaking through these two treatments included field bindweed and smartweed, while lambsquarters had greater than 97.6% control under all treatments on 8 Jul. On 22 Jul all treatments continued to have 90% or greater overall weed control, except metolachlor only and programs with ethofumesate at 0.70 and 0.53 kg ai ha^{-1}. Lambsquarters and smartweed were breaking through ethofumesate at 0.53 kg ai ha^{-1} and redroot pigweed with ethofumesate at 0.70 kg ai ha^{-1}. By 4 Aug pre-emergence programs that contained ethofumesate at 0.87 kg ai ha^{-1} or more maintained greater than 90% weed control of lambquarters and pigweed, while a rate of 1.57 kg ai ha^{-1} or more was needed to maintain 90% weed control or greater for smartweed and field bindweed. If a post-emergence application was added, weed control was 97% or greater for lambsquarters and redroot pigweed and 88% or greater for smartweed and field bindweed.

The prominent weed species in 2017 included common ragweed (*Ambrosia artemissiifolia* L.), lambsquarters, redroot pigweed, galinsoga (*Galinosga parviflora* Cav.), Pennsylvania smartweed (*Polygonum penslyvanicum* L,), purslane (*Portulaca oleracea* L.), and field bindweed (*Convolvulus arvensis*L.). On 10 Jul all programs had greater than 91% overall weed control compared to untreated and the metolachlor only program had significantly less weed control as compared to those programs that also contained ethofumesate at 2.13, 1.57, 1.05 and 0.70 kg ai ha^{-1}. Ethofumasate at 0.87 and 0.53 kg ai ha^{-1} were not significantly different from metolachlor. Nortron at 0.53 kg ai ha^{-1} had reduced lambsquarters control on 10 Jul compared to all other herbicide pre-emergence programs. Control of all other weed species (ragweed, pigweed, galinsoga, smartweed, purslane) was greater than 97% for all other pre-emergence treatments on 10 Jul. By 31 Jul a flush of ragweed occurred in the trial and the only treatments that maintained control were the ones that included clopyralid as a post-emergence application alone or in the 4-way mix. Ethofumesate at 0.53 kg ai ha^{-1} resulted in only 30% lambsquarters control. All other treatments maintained greater than 99% control of pigweed, 92% control of galinsoga, 96% control of purslane; 90% control smartweed, and 81% control of lambsquarters.

HERBICIDE TREATED MULCH. H.M. Mathers*, Mathers Environmental Science Services, LLC, Gahanna, OH (84)

ABSTRACT

Herbicide treated mulches offer a distinct advantage to homeowners and landscapers who wish to decrease their hand-weeding. Herbicide treated mulches have been shown to improve and extend efficacy versus mulches or herbicides applied alone. Mulch Manufacturing, Inc., (MMI), Reynoldsburg, OH has taken the initiative to develop dyed and undyed mulches with herbicides in partnership with Dow AgroSciences for the landscape industry. The MMI products are called Weed² & Feed) Mulch (W^2FM) and contain Gallery® (Isoxaben, Dow AgroSciences, Indianapolis, IN) and Dimension® (Dithiopyr, Dow AgroSciences, Indianapolis, IN). The objective of this study were to compare various bark mulch types (pine wood fiber, hardwood wood fiber, cypress, hardwood and pine), with dye and without dye, treated with Gallery and Dimension, to a competitor mulch (Preen Mulch Plus, Lebanon Seaboard Corporation, Lebanon PA) treated with Gallery and Treflan (Trifluralin, Dow AgroSciences, Indianapolis, IN). In addition, this study evaluated two (W^2FM) colors (red and black) shortly after manufacture (fresh) at two herbicide rates (50 and 150%) compared to red and black (W^2FM) that was produced 2 years ago and stored until the trial initiation. Furthermore, efficacy and phytotoxicity will be compared with the above treatments to a *true check* (no mulch, no herbicide, no seed) (a control check (no mulch, no herbicide and seeded) and three chemical checks of SnapShot 2.5 TG (Isoxaben + Trifluralin, Dow AgroSciences) at 1X and 2X rates (200 and 400 lb/ac, respectively), and FreeHand® 1.75G (Dimethamid-p 0.75% + Pendimethalin 1%) (BASF Corp., Research Triangle Park, NC) and no mulch. Mulches were prepared by Mulch Manufacturing, Inc. and delivered to A. Brown and Sons Nursery, Philipsburg, OH the location of the trial into the custody of Dr. H. Mathers. The trial area was cultivated by tractor immediately before planting on May 23, 2017 and mulched May 24-25, 2017. Six feet was left between rows for bush hog mowing and two ft. between plots was left weedy. The study field had been fallow with no herbicide applications for one year and had a severe pre-existing weed infestation at time of cultivation. Three species of common beddings plants were used for phytotoxicity evaluations including, three Marigold 'Janie Bright Yellow', three Verbena 'Obsession mix' and one Geranium (Pelargonium Xbortorum) per plot. These were planted into bare ground and mulched to a depth of two inches in each plot. Two inches of rain occurred after planting and before mulching and again after mulching and chemical treatments to the three ft. X three ft. plots. At three months after treatment (3MAT) the best efficacy was achieved with the 1X W^2FM Red Fresh with a rating of 9.7 (out of 10, where 10 represents 100% control). This treatment, however, was not statistically different from 1.5X W^2FM Red Fresh (8.7), 1X W^2FM Pine (9.0) and 1.5X W^2FM Black (9.0). There was a significant loss of efficacy with shelf-storage for 2 years, and any mulch that did not contain herbicide. The Preen Mulch was statistically less efficacious at 3 MAT versus the four treatments listed above.

ASSESSING RELATIVE HERBICIDE ODOR WITH A FIELD OLFACTOMETER. S. Askew*, Virginia Tech, Blacksburg, VA (85)

ABSTRACT

Herbicides often emit odors that vary in intensity and olfactory response. A field olfactometer is a device that varies the ratio of carbon-filtered and ambient air to determine a "dilution to threshold" (D/T) value. A low D/T value indicates low odor as small ratios indicate that a large proportion of ambient or nonfiltered air was needed before the odor could be detected. The D/T of several herbicides was determined in a randomized complete block study in Blacksburg, VA to compare relative odor intensity. The study consisted of five treatments and a nontreated check and was replicated in time for a total of four replications. Four herbicide products commonly used for broadleaf weed control in turfgrass and containing equal ratios of 2,4-D, MCPP, and dicamba were compared to a market leading product that contained 2,4-D, MCPP, dicamba, and carfentrazone. Products were all sprayed at 30 GPA using CO_2-pressurized research sprayers at 38 PSI pressure through TeeJet 11004 TTI nozzles. Initial studies were conducted on managed perennial ryegrass turf but the odor of turf in nontreated plots was approximately 45 D/T and this impeded separation of herbicidal odor. Subsequent studies were conducted on pavement where only the nontreated D/T averaged 4.3. At 1 to 2 minutes after treatment, Triad Select and Triplet Low Odor had the lowest D/T of 13.7 and 15.5, respectively. Trimec 992, Triplet SF, and Speedzone had a D/T of 30, 34, and 43, respectively and significantly higher. At 20 to 22 minutes after treatment, all herbicides had between 4 and 15 D/T with few differences. This is the first study to evaluate odor intensity of turfgrass herbicides and the data suggest differences in odor are prevalent.

ANNUAL BLUEGRASS SEEDHEAD SUPPRESSION PROGRAMS FOR FAIRWAY- AND ATHLETIC-HEIGHT TURF. J. Brewer* and S. Askew, Virginia Tech, Blacksburg, VA (86)

ABSTRACT

One of the most problematic weeds in cool-season turfgrass is annual bluegrass (*Poa annua*). It is a winter annual grass that germinates in the fall and is a prolific seedhead producer. Due to the difficulty of controlling annual bluegrass with herbicides, many superintendents revert to managing it to promote better playability and aesthetics on greens and fairways. One factor of management programs is the use of PGRs for seedhead suppression in the spring. Due to the loss of mefluidide (Embark), ethephon (Proxy) is the main product used in suppression programs. The results from Proxy programs can be inconsistent when using common spring only programs especially on higher mown turf areas like fairways. Recent published research has confirmed that early winter applications of Proxy followed by a spring program significantly increases the effectiveness and consistency of seedhead suppression on greens. Unfortunately, there has been very little research conducted to evaluate seedhead suppression on fairways, but seedhead suppression on fairways has been shown to be even more inconsistent than greens. We hypothesized that the same early application programs of Proxy could increase effectiveness and consistency on fairways as well. In 2015 and 2016, multiple studies were initiated to evaluate different Proxy combination programs with products like trinexapac-ethyl (Primo), tebuconazole (Mirage), ethofumesate (Prograss), endothall, prohexanedione calcium (Anuew), flurprimidol plus paclobutrazol plus trinexapac-ethyl (Musketeer), and a few others. These trials were initiated in both Blacksburg, VA, and in the Washington D.C. area, and established on creeping bentgrass and Kentucky bluegrass fairways. The trials were set up as a randomized complete block design with three replications. All trials were sprayed with a CO_2-powered backpack sprayer calibrated to apply 280 L/ha.

In 2015, we observed Proxy plus Musketeer applied early in February followed by a spring program suppressing seedhead production approximately 60%. In another study in 2015 on creeping bentgrass, we observed Proxy programs with Primo, Mirage, and Fiata applied in Nov followed by a spring program reducing seedhead production by 60 to 80% compared to the untreated check. Studies in 2016 evaluated Proxy programs with and without Prograss applied in early December which suppressed annual bluegrass seedheads greater than 60%, while Proxy only programs were significantly lower.

A COMPARISON OF ACCELEROMETER -BASED AND BALL DISPERSION METHODS OF MEASURING GREEN TRUENESS. S. Askew*, J. Brewer, and J. Craft, Virginia Tech, Blacksburg, VA (87)

ABSTRACT

Weeds are among several factors that influence putting green canopy trueness. Recent research elucidated several sources of error that, unless controlled, can undermine determination of putt consistency. By measuring golf ball center of gravity, using a mechanical putter to avoid ball wobble associated with rolling devices, measuring ball directional dispersion prior to near-terminal motion, and brushing green canopies between ball rolls to avoid legacy effects, lateral imprecision of balls rolling on visibly consistent creeping bentgrass or synthetic canopies was constrained to approximately 4 mm m^{-1}. When rolling over an infestation of annual bluegrass, for example, lateral imprecision increased to 9 mm m^{-1}. New methods of measuring ball directional consistency could help determine benchmarks that allow superintendents to monitor canopy consistency or steer agronomic inputs for its improvement. Devices that use accelerometers, like cell phones, can measure multidirectional forces imparted on the rolling ball as it traverses a greens canopy. Two such devices include the Sphero, a virtual-reality gaming and user-interface device in the shape of a large ball and the Parrymeter, a rolling device designed to transfer forces directly from a rolling golf ball into a platform that holds an IPhone or IPod device. We used these two devices to compare correlations between gravitational forces measured by accelerometers within the rolling devices to actual golf ball directional consistency. We also used the Parrymeter to show significant decreases in cumulative gravitational force when rolled over adjacent areas of three bermudagrass putting greens that differed in establishment method.

METHODS TO IMPROVE ESTABLISHMENT OF POLLINATOR PLANTS. M.N. Shock* and S. Askew, Virginia Tech, Blacksburg, VA (88)

ABSTRACT

Initial establishment of beneficial native species is one of the greatest challenges in the institution of pollinator gardens. We are working to develop techniques to target application of activated charcoal around desired seeds at planting. Immediately following this by a spray application of herbicide should improve establishment by reducing weed competition during early development. Within this framework, the interaction of multiple factors is addressed including herbicide mode of action, rate of activated charcoal, method of incorporating activated charcoal, and plant species. The results of these experiments will be applied to future endeavors regarding the design of lower maintenance pollinator supporting native wildflower stands.

GOOSEGRASS (*ELEUSINE INDICA*) SEEDLING EMERGENCE PATTERNS IN TURFGRASS. D.P. Tuck*, K.H. Diehl, and M.T. Elmore, Rutgers University, New Brunswick, NJ (89)

ABSTRACT

Goosegrass (*Eleusine indica*) is a problematic summer annual weed in turfgrass. Economical and effective management programs for goosegrass depend on predicting seedling emergence. The objective of this study was to determine goosegrass seedling emergence patterns in bare ground and cool season turfgrass. Experiments were initiated in April 2017 at the Rutgers Horticultural Research Farm No. 2 in North Brunswick, NJ and the Tamarack Golf Course in East Brunswick, NJ. Plots at the Rutgers site were subjected to three different ground cover treatments; bare ground, perennial ryegrass (PRG; *Lolium perenne*) mowed at 1.25 cm, and PRG mowed at 6.4 cm. Plots measured 1.0 by 1.0 m and goosegrass seedlings were counted and removed on a weekly basis from fixed circles measuring 1000 cm^2 within each plot. Soil temperature and soil volumetric water content were monitored at a depth of 5.0 cm. At the Tamarack location, the experiment was performed on two sites. Plots were maintained as a golf course fairway at a 1.25 cm mowing height. Plots measured 1.0 by 1.0 m and goosegrass seedlings were counted and removed on a weekly basis from fixed circles measuring 500 cm^2 within each plot. Soil temperature was monitored at a depth of 5.0 cm.

At the Rutgers Horticultural Research Farm No. 2 site, goosegrass seedling emergence was first observed on May 26 and last observed on October 20. Ground cover treatment had an effect on the rate of seedling emergence. Season long seedling emergence rates were 8518, 2468, and 1563 seedlings per M^2 in bare soil, 1.25 cm mowing height, and 6.4 cm mowing height, respectively. All treatments had similar patterns of emergence, with the peak rate of germination occurring on June 20, at rates of 270, 51, and 36 seedlings per M^2 for bare, 1.25 cm mowing height, and 6.4 cm mowing height, respectively. In bare soil plots, 65% of all seedlings had emerged by July 4, and 90% by August 4. In fairway mowing height plots, 65% of all seedlings had emerged by July 25, and 90% by August 13. In rough mowing height plots, 65% of all seedlings had emerged by July 15, and 90% by August 12. At both Tamarack Golf Course sites, goosegrass seedling emergence was first observed on May 10 and last observed on September 14. At site 1, the peak rate of seedling emergence occurred on June 3, at a rate of 600 seedlings per M^2. 65% of all seedlings had emerged by June 5, and 90% by June 22. At site 2, the peak rate of seedling emergence occurred on May 20, at a rate of 1817 seedlings per M^2. 65% of all seedlings had emerged by May 25, and 90% by June 4.

COLONIAL BENTGRASS (*AGROSTIS CAPILLARIS*) TOLERANCE TO PRE-EMERGENCE HERBICIDES. M.T. Elmore*, D.P. Tuck, J.A. Murphy, and B.S. Park, Rutgers University, New Brunswick, NJ (90)

ABSTRACT

Field research was conducted in 2016 and 2017 to evaluate the response of colonial bentgrass (*Agrostis capillaris*) to various pre-emergence herbicides. Experiments were conducted on stand of mature colonial bentgrass maintained as a simulated golf course fairway on a sandy loam soil at the Rutgers Horticulture Farm No. 2 in North Brunswick, NJ. Treatments were arranged in a randomized complete block design with four replications and applied to 0.9 by 2.0 m plots using a CO2-powered single nozzle boom with 374 L ha^{-1} of water carrier though an AI9504EVS nozzle. The first applications were made on 14 April 2016 and 9 May 2017. In 2016 colonial bentgrass injury was evaluated on a 1 (i.e., complete injury or death) to 9 (i.e., no injury) scale and turfgrass quality was evaluated on a 1 (i.e., lowest quality) to 9 (i.e., highest quality) scale. In 2017, bentgrass injury was evaluated on a 0 (i.e., no injury) to 100 (i.e., complete injury or death) percent scale. Treatments were evaluated until September in each year.

Treatments consisted of various pre-emergence herbicides applied at one (low rate) and two (high rate) times the maximum registered use rates in creeping bentgrass. Treatments consisted of the following: dithiopyr at 0.56 and 1.1 kg ha^{-1}; pendimethalin at 1.7 and 3.3 kg ha^{-1}; prodiamine at 0.48 and 0.96 kg ha^{-1}; DCPA at 11.8 and 23.6 kg ha^{-1}; dimethenamid-P at 1.1 and 2.2 kg ha^{-1}; bensulide at 11.2 and 22.5 kg ha^{-1}; bensulide + oxadiazon at 6.7 + 1.7 and 13.4 + 3.4 kg ha^{-1}. Treatments were irrigated with 60 mm of water within 4 hours of application.

In 2016, bensulide did not reduce turf quality compared to the non-treated control on any rating date. By 5 weeks after treatment (WAT) bensulide + oxadiazon at 13.4 + 3.4 kg ha^{-1} caused turfgrass quality reductions; these reductions were still apparent at 15 WAT where turfgrass quality was < 3. At 15 WAT, both rates of DCPA also reduced turfgrass quality to < 4. By 20 WAT turfgrass quality was ≥ 5 for all treatments except DCPA at 24 kg ha^{-1} and bensulide + oxadiazon at 13.4 + 3.4 kg ha^{-1}.

In 2017, injury was observed at 3 weeks after treatment for all treatments except low rates of bensulide and bensulide + oxadiazon as well as both rates of prodiamine. By 9 WAT all treatments except both rates of bensulide and the low rates of dimethenamid-P and bensulide + oxadiazon caused colonial bentgrass injury. Both rates of DCPA, the high rate of dithiopyr and pendimethalin caused more than injury than all other treatments (>65%) at 9 WAT. By 20 WAT, < 15% injury was observed in all treatments except the high dithiopyr, pendimethalin, and prodiamine as well as both rates of DCPA. High rates of DCPA and prodiamine cause more injury than all other treatments (>75%).

TEKKEN™: AN SDHI/DMI PREMIX PRODUCT FOR EXTENDED CONTROL OF TURFGRASS DISEASES. B.A. Aynardi*, J.W. Marvin, and A.G. Estes, PBI-Gordon Corporation, Bellefonte, PA (91)

ABSTRACT

[no abstract submitted]

UPDATE ON 2017 WEED SCIENCE RESEARCH IN THE IR-4 ORNAMENTAL HORTICULTURE PROGRAM. C. Palmer*, Y. Lan, J. Baron, and E. Vea, IR-4 Project, Princeton, NJ (92)

ABSTRACT

In 2017, the IR-4 Ornamental Horticulture Research Program sponsored research on two weed science projects: 1) crop safety of over-the-top in-season herbicide applications, and 2) crop safety of over-the-top herbicides applied to ornamental grasses. The goal generally for these over-the-top herbicide applications was to screen herbicides [Biathlon (oxyfluorfen + prodiamine), Dimension 2EW (dithiopyr), F6875 4SC (sulfentrazone + prodiamine), Freehand G (dimethenamid-P + pendimethalin), Gallery SC (isoxaben), Gemini Granular (prodiamine + isoxaben), Pendulum 2G (pendimethalin), SP1770, and Tower EC (dimethenamid-P)] for safety on woody and herbaceous perennials grown in container nurseries. Dimension, Gallery, and Pendulum were also screened on ornamental grasses. Applications were made at dormancy and approximately 6 weeks later for all products. Biathlon was applied to 42 crops; Dimension EW was applied to 13 crops; F6875 4SC was applied to 2 crops; Freehand was tested on 19 crops; Gallery was applied to 10 crops; Gemini Granular was applied to 15 crops; Pendulum G was applied to 6 crops; SP1770 was applied to 9 crops; and Tower was applied on 9 crops. The results from this research will aid in the development of product labels and will help growers and landscape care professionals make more informed product choices.

TOLERANCE OF SEVERAL SEDUM SPECIES TO POSTEMERGENCE HERBICIDES. A. Senesac*, Cornell Cooperative Extension of Suffolk Co., Riverhead, NY (93)

ABSTRACT

A greenhouse trial was conducted to evaluate pre- and post-emergence herbicide tolerance in five sedum species: *Sedum reflexum, Sedum rupestre* 'Angelina', *Sedum album, Phedimus spurius* (*Sedum spurium*) 'Red Carpet' and *Aizopsis ellacombeana* (*Sedum kamtschaticum* 'ellacombeanum').

The species chosen are ones that are popular with commercial growers for producing green roof pallets. The pallets are usually grown outdoors for a few months before shipping and installation. During that time, the flats are susceptible to weed seed invasion. This study was conducted to determine if the sedum species will tolerate spray and granular applications of the following herbicides: topramezone, metsulfuron, and dimethenamid-P + pendimethalin (Freehand). Topramezone (Pylex) is of particular interest because it is a sprayable formulation that can provide post emergence control of some key weed species with over the top applications. Topramezone is currently registered in New York State (including Long Island) for various turf sites. The plants were grown outdoors for several days before treating. The preemergence treatments were irrigated immediately and the postemergence treatments were irrigated 24 hours after treatment. At three days after treatment, the treated flats were moved into a heated greenhouse with 8 hour/day supplemental metal halide lighting. The flats were evaluated for visual signs of injury at 10, 20 and 30 days after treatment.

The results indicate a high level of tolerance of the sedums to topramezone at the 1X and 2X rates. At the 4X rate, only Sedum album showed significant injury to this treatment. These results are promising and suggest that topramezone should be further evaluated.

In general, metsulfuron caused unacceptable injury to all sedum species. No further evaluation of this herbicide seems necessary on these species.

Dimethenamid-P + pendimethalin, which is currently registered for all sedum species, was very well tolerated by all five species at the 4X rate.

HOW DO PRE- AND POST-PLANT WEED CONTROL INTERACT IN ESTABLISHMENT OF PLANTED TREE SEEDLINGS? A.E. Gover*, Penn State, University Park, PA (94)

ABSTRACT

Previous research conducted by this project investigating post-plant weed control effects on tree seedling establishment in unfertilized wildland settings provided results that were barely or not significant. Intensity of pre-plant weed suppression was variable in previous studies, as was planting stock source, and site conditions. In an effort to clarify vegetation management impacts, an experiment was established at two sites evaluating the interaction of pre-plant and post-plant weed suppression. A factorial design with two levels of pre-plant (with or without) and three levels of post-plant (0, 2, 4 years) weed suppression in plantings of tulip poplar (*Liriodendron tulipifera* L.) seedlings was established in riparian settings in Elverson and West Chester, PA. The Elverson site was an Edgemont channery loam (Fine-loamy, mixed, active, mesic Typic Hapludults; NRCS capability class 2e), and the West Chester site was a Codorus silt loam (Fine-loamy, mixed, active, mesic Fluvaquentic Dystrudepts; NRCS capability class 2w). The productivity of the two sites would be inherently similar, but the Elverson site appeared to be more eroded with less topsoil, while the West Chester site was a loamy, alluvial soil. The pre-plant treatment of glyphosate plus sulfometuron plus pendimethalin (2.2 plus 0.023 plus 4.4 kg/ha, respectively) was applied to both sites October 29, 2015, treating 1.8 by 1.8 m squares, arranged in a completely randomized design with eight replications. Containerized seedlings from a single nursery averaging 47 cm in height were planted April 26 and May 6 at Elverson and West Chester, respectively. The post-plant herbicide was applied immediately after planting, using the same mix as the pre-plant treatment. Height of tallest terminal bud and stem diameter was measured on all stems at each site November 10, 2016 and November 1, 2017. Data for 2- and 4-yr post-plant treatments were combined for the analysis. Data were combined in a mixed model analysis, with site as a random variable. There was no interaction between pre- and post-plant treatment for height or diameter for either year. In 2016, a pre-plant treatment resulted in significant increase in height (114 vs. 82 cm) and caliper (11.5 vs. 10.4 mm) compared to no treatment, while post-plant treatment was not significant for either variable. In 2017, there was still a significant effect for total height (162 vs. 133 cm) and caliper (13.4 vs. 11 mm) from pre-plant treatment compared to none, but not for annual 2017 growth (48 vs. 48 cm). Post-plant treated trees showed significantly greater total height (159 vs. 136 cm), 2017 growth (58 vs. 36 cm), and caliper (12.9 vs. 11.5 mm) compared to untreated trees.

INVASIVE VASCULAR PLANT SPECIES AT CUTTYHUNK ISLAND, MASSACHUSETTS, 1932 TO 2017. R. Stalter* and N. Rukhsha, St. John's University, Queens, NY (95)

ABSTRACT

Cuttyhunk Island, Dukes County, Massachusetts USA, comprising 235 ha, 41 25' N, 70 56 W, was formed during the retreat of the Wisconsin Glacier approximately 14,000 years ago. We performed a complete inventory of the vascular flora of Cuttyhunk Island from 30 September 2016 to 6 October 2017 confirming the taxa collected by previous investigators with 26 additional vascular plant species identified in the present study. Of the 364 species, 107 are invasive taxa and 257 are native species. Invasive taxa have increased: 25% of the flora in 1923. 34% in 1974, and 38% in 2017. In numbers of species the largest families in the flora are the Asteraceae (52), Poaceae (37), and Cyperaceae (31). The largest genera are *Carex* (21), *Juncus* (14), *Cyperus* (7) and *Eleocharis* (7). There has been a decline in ferns, orchids, rushes and sedges since Fogg's floristic inventory in 1923. Across the past 93 years of botanical study, the island's flora has changed in response to a dynamic landscape impacted by human activity and coastal storms.

WAVYLEAF BASKETGRASS (OPLISMENUS UNDULATIFOLIUS) IS IN PENNSYLVANIA. NOW WHAT? A.E. Gover* and E.B. Euker, Penn State, University Park, PA (96)

ABSTRACT

Wavyleaf basketgrass (*Oplismenus undulatifolius* (Ard.) P. Beauv.; OPLUN) is a highly shade tolerant, cool-season, stoloniferous, perennial grass first confirmed in the United States in 1996 in Patapsco Valley State Park, Maryland. Subsequently, it was confirmed in Washington D.C. and northern Virginia. A survey in southeast Pennsylvania of Codorus (CODO) and Gifford Pinchot (GIPI) State Parks confirmed presence of OPLUN in disc golf areas in September 2016. A similar survey of these parks in 2009 was negative. Additional surveys of equestrian parking areas at CODO and French Creek State Parks in November 2016 were negative. All Pennsylvania sightings were treated with a glyphosate-based herbicide mixture as they were discovered. 2017 surveys at CODO did not find any additional sites, just additional and surviving plants within the original discovery area. Additional surveys at GIPI did find additional sites. Of particular interest was discovery of well-established patches (10 m diameter) along hiking trails, in addition to the disc golf area findings. A systematic grid-based search at GIPI was initiated in September at GIPI, but was discontinued for fear of dispersing the sticky, ripening seed. Surveys were resumed in November, and apparently have identified the outer edges of the infestation, or at least a discrete infestation.

This leads to the question "Now what?" Possible ends of the response spectrum range from assuming OPLUN is a more impactful invader than existing exotics and worthy of diverting effort to survey and containment. Our initial efforts align with this, as we have been surveying and treating in low priority sites already significantly invaded with multiple species. A scenario at the other extreme is regarding OPLUN as just another invader, and limiting suppression efforts to higher-priority land units. OPLUN is moving into a region already heavily invaded by Japanese stiltgrass (*Microstegium vimineum* (Trin.) A. Camus), which apparently fills a very similar niche as OPLUN. Ecologically, is this a new threat?

OPLUN is now a regulated plant in PA, listed as a "Class A" Noxious Weed, accompanied by the language "Preventing new infestations and eradicating existing infestations of noxious weeds in this classis high priority." In an effort to facilitate wider consideration, an informal working group has been established, defined by participation in a OPLUN-specific listserve. The listserve is open to all, and initial membership is comprised of public-access land managers, regulators, and academics. Short-term, survey and management will be limited to in-house or local collaborative efforts, and education and outreach efforts to alert landowners and outdoor enthusiasts of this new threat. With the majority of forestland in PA being private, and OPLUN largely limited to shaded woodland settings, best case is that OPLUN will be effectively suppressed on public lands, while a segment of private lands will continue to serve as refuge and reservoir for continued spread.

SYMPOSIUM: INTEGRATING NEW TECHNOLOGY TO MEET THE FUTURE CHALLENGES OF AGRICULTURE: A SHARING OF EXPERIENCES.

ABSTRACT

The Northeast Plant, Pest, and Soils Conference (NEPPSC), with Participating Societies that include: the Northeastern Regional Branch of American Society of Agronomy, Crop Science Society of America, Soil Science Society of America American Society of Horticultural Science - Northeast Division and the Northeastern Weed Science Society.

The symposium session focused on how the respective societies can continue to evolve, and adapt and respond to the needs of the US Crop Production, Turf, and Forestry industries. Moreover consider how new technological developments, resulting from the hard work of the joint society members, can be rapidly integrated into productive practices resulting in uninterrupted food supply and global trade. The symposia shared ideas and experiences across our NEPPSC societies. Noting that the applied scientists will remain increasingly relevant to agriculture and critical to meeting the challenge of feeding the growing population in a sustainable manner.

After a brief background presentation by the moderator, Rob Hedberg USDA-NIFA, National Director for the Sustainable Agriculture Research and Education Program and the National Program Leader for the Minor Use and Specialty Crop Pesticides Program (IR-4), provided an overview of their work, with an emphasis on Tactical Sciences for Protection of the U.S. Agricultural Enterprise.

Then Dr. Norm Leppla, University of Florida, Professor & Program Director for IPM, spoke about the CAST paper he published with Dr. Susan Ratcliffe et al. on Crop Protection Contributions toward Agricultural Productivity--A paper in the series on The Need for Agricultural Innovation to Sustainably Feed the World by 2050. See below for the abstract of this presentation.

Dr. Nicole Juba, from Syngenta Crop Protection, provided a more in-depth overview of plant genetics, including CRISPER techniques and RNAi advances and application.

Mr. Chip Bowling, a corn grower in Newburg, MD., and past President of the National Corn Growers Association, finished the plenary session with a presentation on the Future Challenges for Agriculture.

Outcome from the Breakout sessions.

Questions Discussed and Responses from the participants

1. **RNAi Technology – Discuss some of potential applications of RNAi and how it may assist Agriculture and pest control.**
   Control pest (insects), reduce/remove potential for resistance.

Responses…

Potential Applications.

1. All areas of pest control and resistance management
   a. Insect
      i. Agronomic as well as Public Health (bed bugs/mosquitoes)
   b. Virus
   c. Nematode
   d. Fungal
   e. Rust
   f. Weed Control/Resistance

Benefits.

1. Eliminate/reduce use of other products that are more hazardous/problematic (e.g. methyl bromide)
2. Topical applications allow for flexibility of use.
3. Can be used to "turn off" undesirable process (e.g. browning or production of allergy causing proteins)
4. Can be produced quickly relative to other technologies.
5. Softer on beneficial insects, very targeted application and sites of action

Concerns/Issues

1. Need to understand your pathways/basic research of processes
2. Resistance development
3. Is all RNAi created equal? Can insects/pests overcome RNAi?
4. How fast does resistance occur?
5. Does RNAi negatively impact desired characteristics/process
6. Lots of unknowns, need more data.
7. What is the "shelf life"?

2. **Cover crops – Share current uses and how the use of cover crops may further evolve.**
    a. **Considerations for growers using cover crops to prepare for extreme weather events, such as storms, excessive rain, seed predators, etc. and ensure success?**
    b. **How use of cover crops can be used to manage herbicide resistance?**
    c. **Relationships with herbicides and cover crops. How best to complement each other?**
    d. **As noted above, what maybe some of the best mixtures to use.**
    e. **Looking at the win, win, win with predictability and economics.**

Responses…

Considerations and Benefits.

1. Extreme weather
    a. Erosion control
    b. Water infiltration, shorter time to get back to work (drivability)
    c. Dry out soils, can create problems for establishing cash crops
2. Soil Health – vital to long term payoff
    a. Greater flexibility of crops
    b. Nutrient management
3. Brings new experts to the table, spreads the work and greater collaborations and innovation. More year round work, not just busy seasons.

4. Manage Herbicide Resistance
    a. Smother weeds
    b. Allelopathy
    c. Herbicide usage – does it decrease?
    d. Fewer weeds – smaller weeds more effective, less exposure
    e. Manage Cover crops to avoid them from becoming weeds. May need herbicides to kill/control cover crops – esp. with mixtures.
    f. Herbicide carry over may damage cover crops
    g. Choose timing wisely in light of crop
5. Mixtures and management
    a. Lack of varieties to choose from – lack of information on traits
    b. VNS is most popular
    c. Best mixtures depends on the use/objectives and region
    d. Consistent traits can be a problem.
    e. Select herbicides and times to kill cover crop
    f. Consider or allows flexibility to integrate livestock into rotations.
6. Incentives are needed to encourage growers to try new methods
    a. Word of mouth, tried and true, reliable

3. **Please provide your experience with an invasive pest. What cropping system was impacted and how extensive was the issue resulting from the invasive pest? How did it influence existing IPM programs?**
    a. **How best to respond – training and best means to inform to "activate the system" to respond.**
    b. **What is the best means to effectively share the information in real time, building partnerships and maintaining infrastructure.**

Responses…

Examples and responses:

1. Jointhead grass and certain other invasive grasses were becoming problematic in pastures in the region with no viable control options. The mid-Atlantic weed group came together and reached out to industry for possible solutions. Opened up residual options for grass control, Prowl was registered as a result.
    a. Training – pesticide applicator recertification, fact sheets, extension meetings etc.
    b. Sharing of information – weed scientists carried out field experiments on a concerted basis and shared information at NEPPSC meeting.
2. Japanese Knotweed. Was found in the roadsides. Some states are working closely with DOT and Conservation agencies. Biocontrol efforts being sought (*Aphalara itadori, Ostrinia ovalipennis,* and *Gallerucida bifasciata*). Communication is key.
3. Kudzu biocontrol – Some fungus infests Kudzu in SE Virginia. *Beauveria bassiana* has been an effective control measure
4. Multiflora Rose – small ruminants (goats) selective feed promoting growth and expansion. Mixed grazing encouraged.
5. Palmer Amaranth – Native to the desert of the SW US. Scientists have been talking about it for more than 10 years. It will continue to be a problem and we will continue getting questions. Diversify programs to reduce selection pressure on other weeds
6. Spotted lanternfly – easy to control, but challenge is getting to it.
7. Snapping turtle
8. Wild Parsnip
9. Giant Hogweed – found in Canada and many New England states. Moving south. Concern to homeowners – mistaken for cow parsnip and sometimes intentionally planted.
10. Brown marmorated sting bug. Monitoring sprays. It has spread to other regions from Mid-Atlantic region where the initial problem was.
11. Soybean rust- used to be a major concern.
12. Spotted wing drosophila – another emerging insect pest concern.

Training and Communications

Early detection, rapid response and education are all very important.

Education needs to also involve the general public.

Citizen science programs with Master Gardeners and 4-H youth being considered in some states to assist with monitoring/scouting.

Work with growers; Get farmers involved in delivery of solutions and use economics to educate, may help to reach the unreachable.

It is important to have proper identification. Geo-mapping, photos and long-term storage of data "bugwood"

There are some mapping/tracking programs for specific pests. PIPE- late blight, cucumber mosaic virus, CDM, Scouting, Sentinel plots.

Other Considerations.

- Personal information maybe a roadblock, a grower may not want to share their problems.
- Community based information via extension
- Government/Industry/academic/Extension – working together to monitor, detect, and react.
- Certification of weed-free crops and equipment.
- Look toward other systems used – for disease and insect control.
- Share data.

4. How do you see the use of drones to benefit agriculture? Discuss some examples where they are already being used and further potential for these tools.

Mapping weeds, dispersing beneficial insects.

Precision spraying or smart sprayers. Please share some of your experiences with "camera assisted" sprayers and their application. For example, lettuce thinning in California and selective weed spraying in crop production.

Responses…

1. Applications
   a. Monitor and scout pests and incidence of invasive pests
   b. Detect crop stresses
   c. Make spray applications
   d. Already being used in many situations – turf, disperse beneficials, mosquito sprays
   e. Useful on rough terrain
   f. Can be used to map fields
   g. Some products for weed control

2. Benefits - can save on labor

3. Concerns/Issues – Air traffic control

CROP PROTECTION CONTRIBUTIONS TOWARD AGRICULTURAL PRODUCTIVITY. N.C. Leppla, Professor and Program Director, IPM, University of Florida, IFAS, Gainesville, FL

ABSTRACT

The current plant protection revolution is driven by the biological realities of pesticide resistance, various market forces, and real or perceived side effects of pesticides. Crop protection chemicals have been miraculous, but their automatic use is no longer efficacious or justifiable. Integrated pest management is the preferred approach, and pest prevention is a key component in its success. Current and future methods used to control plant diseases, insects, and weeds will be described, along with the need for scientists from all the pest management disciplines to develop and deliver integrated strategies for managing pests while preserving ecosystem services and farm productivity.

NEWSS YEAR-END FINANCIAL STATEMENT 2017

AUDIT REPORT – Kurt Vollmer

Financial documents and bank accounts were audited by,
Todd Davis and Quintin Johnson on December 11, 2017.

BEGINNING STATEMENT: As of October 31, 2016, the NEWSS checking account showed a balance of **$53,502.86**, while the savings account balance was **$31,844.57**. The endowment fund balance was **$23,614.39**

ENDING STATEMENT: As of October 31, 2017, the NEWSS checking account showed a balance of **$65,956.35**, while the savings account balance was **$31,850.92**. The endowment fund balance was **$26,194.01**

Net gain in assets for the fiscal year: $124,001.28 - $108,961.82 = $15,039.46

FISCAL YEAR ACCOUNTING

New income as shown on statements for November 2016 through October 2017 including earned interest and endowment gain was **$62,915.17**

New deductions from checking for November 2016 through October 2017 totaled $**47,875.71**

Net gain** in assets: **$62,915.17 - $47,875.71 = 15,039.46

Non Annual Meeting Revenue/Expenses:

Non annual meeting revenue, which was comprised of interest earned as well as member dues only, weed contest support from sustaining members, and other contributors ($6,500.00) totaled **$9,335.97**

Non annual meeting expenses (Dishonesty bond, website, weed contest, administrative costs, site visits, Dir. Science Policy (2017) and CAST dues) totaled **$28,102.12**

Annual Meeting Revenue/Expenses:

Annual Meeting revenue (registration, dues, BASF award, industry social and sustaining dues, proceedings residuals, meeting planning) totaled **$49,896.70**

Annual meeting expenses (hotel + AV, PayPal fees, programs, awards, registration desk assistance, retired fellow reimbursement, and DSP travel to annual meeting) totaled **$19,773.79**

Education Fund Contributions

Contributions received for the Education Fund totaled **$3,682.50** for the 2017 fiscal year.

NEWSS FISCAL STATEMENT

NOVEMBER 1, 2016- OCTOBER 31, 2017

Income Received	Amount
Joint Meeting Total Registration	$32,478.50
Member Dues Only	$250.00
BASF awards	$300.00
Industry Social (3 Sponsors at $3000.00)	$9,000.00
Interest earned on savings account	$6.35
Net gain in endowment	$2,579.62
Sustaining Membership: 15 to date	$7,750.00
Sustaining Designated for Weed Contest	$4,000.00
Additional Weed Contest Support	$2,500.00
Endowment Contributions	$3,682.50
Proceedings Residuals	$118.20
NEPPSC Planning Contribution	$250.00
Total NEWSS Income Nov 1, 2016- October 31, 2017	**$62,915.17**

Expenses Paid	Amount
PayPal Fees	$1,190.54
Sheraton Society Hill October 2016 Meeting	$194.40
Joint Meeting Hotel + AV Cost	$15,046.94
Alphagraphics - Mtg Programs	$585.71
NEPPSC Mtg Registration Reimbursement Retired Fellows	$200.00
NEPPSC Registration Desk Assistance	$821.00
Lee Van Wychen NEPPSC Travel Costs	$171.00
EC Dishonesty Bond Insurance Renewal	$359.00
Website "A Small Orange" domain fee for NEWSS org.	$200.00
New NEWSS/NEPPSC website	$5,000.00
Administrative	$516.23
Lewis Body Dementia: Henry Wilson Memorial	$50.00
Annual meeting Membership Awards - Etch Art	$539.00
Annual Meeting Poster Awards (1st and 2nd)	$300.00
Annual Meeting Student Paper Awards (1st and 2nd place)	$300.00
Annual Meeting Photo Contest (1st, 2nd and 3rd place)	$175.00

Collegiate Weed Contest host fee	$17,976.89
WSSA Dir. Science Policy YEAR 2017 Dues	$2,500.00
CAST dues	$1,500.00
2018 Planning Meeting	$250.00
Total NEWSS Expenses paid Nov 1, 2016- October 31, 2017	**$47,875.71**

Total Fiscal Year Summary

Total Income	$62,915.17
Total Expenses	$47,875.71
Net Gain	**$15,039.46**

NEWSS Account Balances October 31, 2017

Checking	$65,956.35
Savings	$31,850.92
Education Fund	$26,194.01
Total Net Assets	**$124,001.28**

NORTHEASTERN WEED SCIENCE SOCIETY

2018 NEWSS BUSINESS MEETING

Call to Order

President Randy Prostak called the meeting to order at 1637.

President's Comments

President Prostak waxed nostalgic, and shared a few only-moderately inappropriate anecdotes.

Offered kudos to Kate Venner, Kurt Vollmer, Sudeep Matthew, and Quintin Johnson for their efforts upgrading the Society website.

Reminded the Society to share input on the ongoing status of NEPPSC.

Student Endowment Fund close to $33,000. Interest revenue was tapped for 2018 Quiz Bowl awards.

Reminded us that he is not stable, not a genius. But we love him anyway.

Praised the Society's can-do attitude, and helpfulness, in general, and specifically during his tenure in the Presidential arc.

Encouraged members to serve on the Board.

Acceptance of the 2017 Business Meeting Notes

A few minutes were taken to review the minutes from the 2017 Business Meeting. Jeff Derr moved to approve, Dave Yarborough seconded. Minutes approved without revision.

Membership Chair Report - Sudeep Matthew

MEMBERSHIP

Membership report distributed, with 10-year summary. After a peak of 140 Regular Members in 2016, numbers decreased to 96 then 65 in 2017 and 2018 respectively. Sudeep reminded us that it is up to us as members to encourage other members, and that small numbers matter, as they add up in a Society of our size.

NECROLOGY

Henry P. Wilson, 1/7/17. Society President 1975-76.

Treasurer's Report - Kurt Vollmer

Kurt reviewed submitted report.

Audit was conducted 12/11/17 by Todd Davis and Quintin Johnson. The accounts were found to be in order.

Education Fund was at $26,194.01 as of October 31, but closer to $35,000 now.
- the first use of revenue, from interest, was Amazon gift cards for 2018 Quiz Bowl winners.
- cash prizes are being considered for 2018 Weed Contest winners.

BASF is hosting 2018 Weed Contest.

Encouraged Society input on usage of NEPPSC meeting funds.

Awards Committee - Shawn Askew

RECAP OF 2018 SOCIETY AWARDS

GRADUATE STUDENT PRESENTATION AWARDS – Rakesh Chandran

Review of comments - overall very good this year; "Tell your story", limit details to the essentials, don't lose your audience in the details, clearly state objective(s) and hypothesis; readable font size; uncluttered charts and graphics; limit scope of conclusions to the data provided; be clear about statistics and experimental design; preliminary explanation of graph/chart set-up; use SI units.

Shawn adds - only share information relevant to the particular experiments; introduction should be complete within 5-6 minutes.

GRADUATE STUDENT POSTER AWARDS - Matt Elmore

Review of comments - overall quality was good this year; follow the rubric on the website; poster should be stand-alone; make sure figures are needed; proofread, proofread, proofread.

PHOTO CONTEST - Art Gover

There were 22 total entries from 6 contestants. Art reminded the Society that the Photo Contest is an easy way for members to be in the running for real money, doing something they do as a matter of course.

Weed Science Policy - Lee Van Wychen

Lee provided a quick update (yes, really) on the following topics:

- WSSA EPA Liaison - Greg Kruger has assumed this role, taking over for Mike Barrett.
- USDA-NIFA Fellow - Donn Schilling is stepping down after three years of service. WSSA will need to appoint his successor.
- USDA Leadership Vacancies - Lee reported there is an effort to get Rich Bonanno appointed to the role of Under Secretary for Research, Education, and Economics; a position regarded as USDA's chief scientist.
- FY 2018 Federal Budget - we are currently working under a Continuing Resolution through January 19, with many issues unresolved in working towards an actual budget. Ag appropriations are at satisfactory levels, provided sequestration caps are not implemented.

- Dicamba - EPA announced label changes to the dicamba products Extendimax, Engenia, and Fexapen on October 13, 2017. EPA has indicated that if the issues encountered in 2017 are repeated, 2019 registration is questionable.
- Glyphosate - WSSA in on record stating "the IARC review process for glyphosate was flawed and represents a case of gross scientific negligence. There is no question that IARC arrived at their conclusion due to their inclusion of the positive findings from a selection of studies with known limitations, a lack of reproducible positive findings, and the omission of the negative findings from credible and reliable research." Lee also summarized the November 9, 2017 results in the Agricultural Health Study published by the National Cancer Institute; and EPA's December 19, 2017, human health draft risk assessment.
- EPA Herbicide Resistance Guidance - Referred to as Pesticide Registration Notice 2017-2, this document will apply to all herbicide uses except residential settings (lawns).
- National Invasive Species Awareness Week (NISAW) - February 26 - March 2, 2018.
- Farm Bill - Congress will begin working on the updated Farm Bill in 2018.
- 2018 Weed Survey - focus will be on weeds in non-crop areas.

CAST Representative Report - Carroll Moseley (for Rob Richardson)

Summarized 2017 publications, and topics in the proposal and development stage, and new leadership of the Plant Working Group (compiled by Rob Richardson).

Officer Changeover and Presentation of the NEWSS Gavel

Randy Prostak relinquished the NEWSS presidency to Carroll Moseley, and presented The NEWSS Gavel.

Carroll's first order of business was to present the ceremonial plaque and gavel to now Past-president Randy Prostak.

New Business - Carroll Moseley

NOMINATION COMMITTEE

The Committee offered Art Gover of Penn State as candidate for Vice-president. President Moseley opened the floor for additional nominations. Jeff Derr moved and Dave Yarborough seconded that nominations be closed. Art Gover approved as new Vice President.

Carroll Moseley appointed Randy Prostak and Jeff Derr to the 2018 committee. Randy Prostak nominated Sudeep Matthew. Michael Flessner nominated Shawn Askew. Anthony Witcher nominated Thierry Besançon. Approved by acclamation.

Jerry Baron, Matthew Elmore, and Matt Cutulle were nominated to Resolutions Committee.

Membership Chair - Stacie Pickard nominated, approved by acclamation.

WSSA Representative - Rakesh Chandran was appointed.

OTHER BUSINESS

Society Editor Michael Flessner reminds us that the abstract submission feature on the wssaabstracts website is now open for delinquent abstracts.

Next year's meeting dates and venues are not yet final, but will be announced within 30 days.

Collegiate Weed Contest is at Pine Level, NC, hosted by BASF, July 23-25, 2018.

Carroll reminded us that NEWSS is an exemplary setting for graduate student development and camaraderie, and comparatively unique compared to other societies.

Presentation of the NEWSS Board

The board was presented, with group photograph to follow adjournment.

Adjourn

Shawn Askew moved, Sudeep Matthew seconded, motion approved at 1801.

Author Index

Northeastern Weed Science Society (NEWSS)
P.O. Box 25
Woodstown, NJ 08098
USA

ISBN 978-1-5108-6295-1

CURRAN ASSOCIATES INC.
proceedings.com